ANCIENT

THE MEDITERRANEAN W

THE MEDITERRANEAN	
	…lithic Monuments Brittany, Spain, Britain
Cycladic Sculpture Spread of Aegean Culture Knossos - Crete	INDUS VALLEY CULTURE in India c.2500-1650 B.C. Figures in bronze, terra cotta
Mycenae	
Primitive Geometric Art in Greece - Statuettes	
ARCHAIC GREEK SCULPTURE Etruscan Sculpture	Hispano-Phoenician Sculpture in Iberia
CLASSICAL PERIOD Parthenon - Phidias HELLENISTIC PERIOD Victory of Samothrace - c.200 Laocoon c. 25 B.C.	Hindu Sculpture Alexander in India - Influence of Greek Art
Roman copies of Greek works Portrait busts-monuments	Provincial Roman Sculpture

THE OBSERVER'S
POCKET SERIES

• • •

THE OBSERVER'S BOOK OF SCULPTURE

The Observer's Books

BIRDS
BUTTERFLIES
WILD ANIMALS
GRASSES
HORSES AND PONIES
AIRCRAFT
ARCHITECTURE
SHIPS
COMMON INSECTS
COMMON FUNGI
AUTOMOBILES
RAILWAY LOCOMOTIVES
GARDEN FLOWERS
CACTI
FLAGS
SEA AND SEASHORE
LICHENS
FURNITURE
WILD FLOWERS
TREES
FRESHWATER FISHES
DOGS
GEOLOGY
FERNS
LARGER MOTHS
MUSIC
BIRDS' EGGS
MOSSES
WEATHER
POND LIFE
PAINTING
SEA FISHES
CATS
ASTRONOMY
MODERN ART
CHURCHES
SCULPTURE

Also by William Gaunt

THE OBSERVER'S BOOK OF PAINTING
THE OBSERVER'S BOOK OF MODERN ART

THE OBSERVER'S BOOK OF

SCULPTURE

By
WILLIAM GAUNT

With 8 plates in full colour and 62 monochrome reproductions from photographs

FREDERICK WARNE & CO. LTD.
FREDERICK WARNE & CO. INC.
LONDON · NEW YORK

LONDON ENGLAND
1966

LIBRARY OF CONGRESS CATALOG
CARD NO. 66–15684

Printed in Great Britain by
Cox & Wyman Ltd., London, Reading and Fakenham
1805. 1065

CONTENTS

LIST OF ILLUSTRATIONS

Plate

Plate

ACKNOWLEDGMENTS

THE author and publishers wish to acknowledge the kind assistance given by the various museums, galleries, owners and agents in connection with the illustrations in this book.

Pl. 41 is reproduced by gracious permission of Her Majesty The Queen; Pls. 4, 12, 14, 16, 17, 18, 21, 25 and 40 are reproduced by courtesy of the British Museum; Pls. 20, 22, 23, 24, 26, 27, 29, 30, 31, 39, 42, 43, 46, 47, 48, 50 are reproduced from photographs from the Mansell Collection; Pls. 6, 7, 8, 9 and 11 are reproductions from the book *The Egyptian Museum, Cairo*, published by Kümmerly and Frey, Berne; Pls. 33 and 52 by courtesy of The Victoria and Albert Museum; Pls. 53 (above) and 54 by courtesy of the Trustees of The Tate Gallery, London; Pls. 13, 28 and 32 by courtesy of the Louvre, Paris. Thanks are also due to the Museum d'Histoire Naturelle de Toulouse for Pl. 1 (photograph by Yan); to Musée de St Germaine-en-Laye, France, for Pl. 2; The Information Bureau, India House, London, for Pl. 3;

Musée de l'Homme, Paris, for Pl. 5; The Arts Council of Great Britain for Pls. 10 and 63; Museo Nacional de Antropologia, Mexico, for Pl. 15; The Birmingham Museum and Art Gallery for Pl. 19; The Courtauld Institute of Art, London, for Pls. 34 and 35; The Gordon Fraser Gallery for Pl. 36, photographed by Kerry Dundas; The Warburg Institute and National Monuments Record, London, for Pl. 37; The Museum of Fine Arts, Budapest, for Pl. 38; The Royal Academy of Arts, London, for Pl. 44; Museo d'Arte Antica, Milan, for Pl. 45; The Ashmolean Museum, Oxford, for Pl. 49 (above); Mallett at Bourdon House Ltd., London, for Pl. 49 (below); A. F. Kersting for the photograph reproduced on Pl. 51; Les Archives photographiques, Paris, for Pl. 53 (below); Le Musée Rodin, Paris, for Pl. 55; Marlborough Fine Art Ltd, London, and Henry Moore for Pl. 56; Marlborough Fine Art Ltd, London, for Pl. 61; John Goldblatt for the photograph reproduced on Pl. 57; Henry Moore for Pl. 58; A. Ronald Traube for the photograph reproduced on Pl. 59; The Museum of Modern Art, New York, for Pl. 60; ARTED, Editions d'Art, Paris, for Pl. 62; and General Motors Corporation, Detroit, for Pl. 64.

While every effort has been made to trace the present copyright owners of works reproduced, the publishers wish to apologise in advance for any inadvertent failure to make suitable acknowledgment.

PREFACE

THERE can be few visitors to a great centre of art or museum who have not been thrilled by some work of sculpture, for sculpture contains the whole of human history. Such works tell of wonderful civilizations in the past, as the horsemen of the Parthenon frieze in the Elgin Room in the British Museum and the Winged Victory in the Louvre tell of the glory that was Greece. Sometimes they are the only record that is left of a race unknown to us otherwise, like the strange carved figures of Easter Island (Plate 4). From prehistoric times to the present day, sculpture has reflected every stage of progress and thought.

The panorama of art and history thus presented is the theme of this book. A companion to the two books on painting in the *Observer's* series by the same writer, it is an attempt to outline the story of sculpture in similar fashion, to give an idea of the motives that have called it into being to describe the different forms it takes and how it may be viewed as an aid to historical study and as an art with its own special beauties.

In his handling of solid materials the sculptor has to deal with more problems of craftsmanship than other artists. Technical processes which play an important part in his work are briefly described, but only so far as may add to the observer's appreciation of the finished result. A glossary of technical or specialized terms is given. Many masterpieces of the past were produced by artist-craftsmen whom we do not know by name or as individuals. Their work is discussed in relation to the periods in which they lived but a biographical section on the masters of more recent times who are distinct in name, personality and achievement is included.

Sculpture is not only the record and product of vanished cultures, it is a living art in the twentieth century. It has taken new forms and gained a renewed vigour in response to the changing conditions of the age. To trace its connection at the present time with the work of earlier periods and to follow its fresh departures and consider their meaning has therefore seemed essential.

The plates have been chosen to illustrate sculpture through the ages by some of the most typical examples of the great epochs and great artists. They are not confined to works in museums but include those remaining on the site for which they were intended and offering to the city-dweller and traveller as rich a source of pleasure and interest as the famous buildings of which they are so often a part. There are instances, also, of sculpture in private collections and adapted to a domestic setting.

Thanks are due to all the public authorities and private owners by whose courtesy the reproductions have been made.

W. G.

CHAPTER I

THE NATURE AND PURPOSE OF SCULPTURE

THE word "sculpture" derives from the Latin verb, *sculpere*, meaning "to carve, hew or cut", though it has a wider application than to carving alone. It is generally understood to refer also to the practice of moulding some pliable material such as clay or wax and the Greek word for "that which can be moulded", πλαστικος, provides an alternative description of sculpture as "the plastic art". It can in fact be any means of giving an intended shape to formless matter.

Both carving and modelling were practised for tens of thousands of years before the Greeks and Romans provided the terms we still use. They were arts of prehistoric man and as such answered to some basic human need. Like painting, sculpture had its origin among the cave-folk of the old stone age, the first artists of all. How they came to develop this artistic faculty is a question answered partly by what seems likely in the circumstances and partly by actual evidence.

It is probable that in the first instance the cave man felt the simple delight in handling substance that children show at the seaside when they play about with wet sand. In the caves of the Dordogne in south-west France where so many amazing discoveries of prehistoric art have been made, there were, and are, deposits of soft clay in which the impress of prehistoric hands has been found. Some were just accidental. An explorer of modern times pulled himself round a narrow turning in one curve by taking a grip on a projection in the wall. Examining it later he found under the hard thin glaze left by time, handprints indicating that a palaeolithic man had negotiated the turn in exactly the same way!

Yet other prints show an interest in the marks of the fingers by those who left them. Series of parallel lines, straight or curved, were dragged through the clay, a first step towards geometrical design. A later stage shows the exercise of the imitative faculty natural to human beings.

The uneven surfaces of cave wall and floor, the varied forms of stalactites and stalagmites seem to have incited the prehistoric artist to shape them further into the likeness of the animals they knew. A stalagmite, for instance, at Font-de-Gaume in the Dordogne suggested the tail and hind legs of a horse and the work of the hand completed the suggestions of nature.

One might suppose, then, that sculpture began as a kind of play and that it developed with a growing delight in the exercise of a skill which eventually produced representations of animals, between 20,000 and 10,000 years ago, unique in realism and vigour. Most authorities, however, have concluded that there was a more practical aim behind them. They were magical images, that is, thought to give their creators power over what they represented, especially the animals which they hunted. The cave paintings point to this conclusion even more definitely than sculpture, when they show arrows affixed to some vulnerable spot in the quarry depicted.

There have been many traces in the past of the primitive belief that the representation of an object becomes identified with the object itself. It is part of the legend of witchcraft that a small figure modelled in wax would enable the modeller with malicious intent to harm or kill the person it portrayed by sticking pins into it, the assumption being that the living person would suffer a corresponding wound. There are actual records of these harmfully intended attempts at magic in comparatively recent times. Even today there are primitive peoples who shrink with terror from the sight of the artist sketching, in the belief that their portrait in the hands of others would place them at their mercy. A figure in three dimensions, and therefore the closer to reality, might be imagined as still more powerful in this respect than an image on a flat surface.

The world of the prehistoric hunter was ruled by animals, either dangerous or of value as food, and the number of their animal images, drawn, painted or sculptured may be looked on as inspired by this superstitious idea of how they might gain command over their environment. Yet different problems confronted the human race at the next stage of its progress. When groups settled in one place and began to till the soil, sow the seed and reap crops they became aware of greater and more mysterious powers than the mammoth or cave bear. The kindly influence of the sun, the destructive powers of storm, fire and flood had a more direct bearing on the means of livelihood of a primitive agricultural community.

The need was felt to define and propitiate these forces. Sculpture made them visible. With the aid of various rites, ceremonies and tributes paid to an idol, it was supposed that the kindly attention of the power it represented could be enlisted. This was a different matter from directing what is called "sympathetic magic" against bison or deer. To express the idea of the superhuman, the attributes most admired or dreaded were called upon. To this end human and animal qualities of ferocious or awe-inspiring appearance and emblematic of physical strength were often combined.

In this attempt to give shape to unknown forces we can see one reason why the marvellously skilled realism of animal art among the prehistoric hunters eventually disappeared to be replaced by images of a more mysterious kind. These were mysterious because in fact they represented mysteries. Now began that development which throughout the world was to link sculpture with religious belief. In ancient Europe, in Mesopotamia, in Egypt and the African continent, in India and China, in Polynesia, in America before the arrival of Columbus, the motive was essentially the same, though sculpture varies with the degree and kind of civilization attained.

The growth of cities in ancient times added to the importance of sculpture. Organized government

1. PREHISTORIC: *Bisons modelled in clay, Tuc d'Audoubert Cave, France*

2. PREHISTORIC: *Head of a Girl in Mammoth Ivory*

(*Musée de St Germain-en-Laye, France*)

3. PRE-HISTORIC: *Nude Dancing Girl, Mohenjo-Daro. Bronze.* c. 2500 B.C.

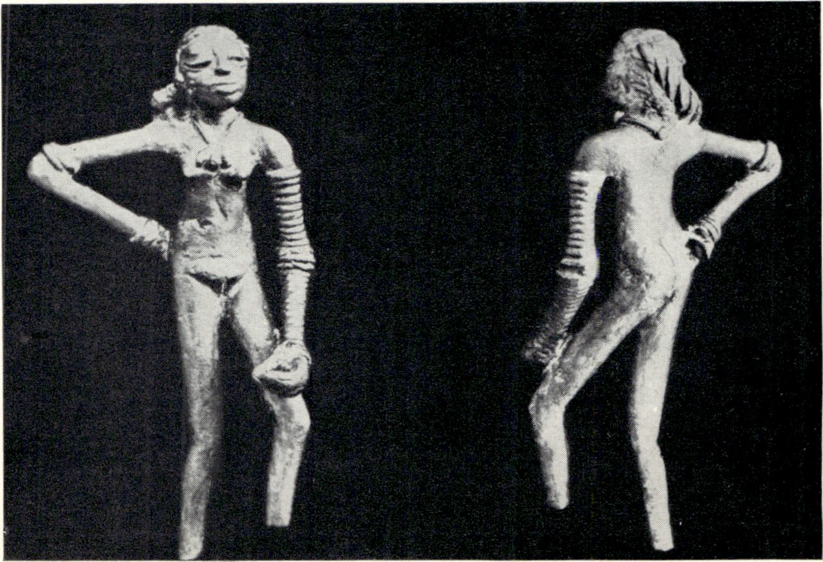

(*Delhi Museum*)

4. PRIMITIVE: *Easter Island Figure*
(*British Museum*)

5. PRIMITIVE: *New Ireland Mask*
(*Musée de l'Homme, Paris*)

6. EGYPTIAN: "*Sheikh El-Beled*" (*Ka-Aper*).
Sycamore wood. Fifth dynasty.
(*Cairo Museum*)

7. EGYPTIAN: *Seated Scribe with Papyrus Scroll. Limestone* c. 2500 *B.C.*
(*Cairo Museum*)

8. EGYPTIAN: *Unfinished Quartzite Head of Nefertiti. Eighteenth dynasty.*
(*Cairo Museum*)

9. EGYPTIAN: *Coffin Cover of Queen Ma'et Kere*
(Cairo Museum)

10. EGYPTIAN: (above) *Unguent Spoon in the form of a duck, held by a swimming girl.* c. *1400 B.C.*

(*Cairo Museum*)

(below) *Woman Grinding Corn.* c. *2400 B.C.*

11. EGYPTIAN: *King Mycerinus between the Goddess Hathor and a Provincial Goddess. Green slate. Fourth dynasty.*
(*Cairo Museum*)

12. ASSYRIAN: *Lion Hunt of Ashur-bani-pal. Seventh century* B.C.
(*British Museum*)

13. PERSIAN:
Coloured Ceramic Frieze of Archers, Persepolis. Sixth Century B.C.

(Louvre)

14. AZTEC: *Crystal Skull*

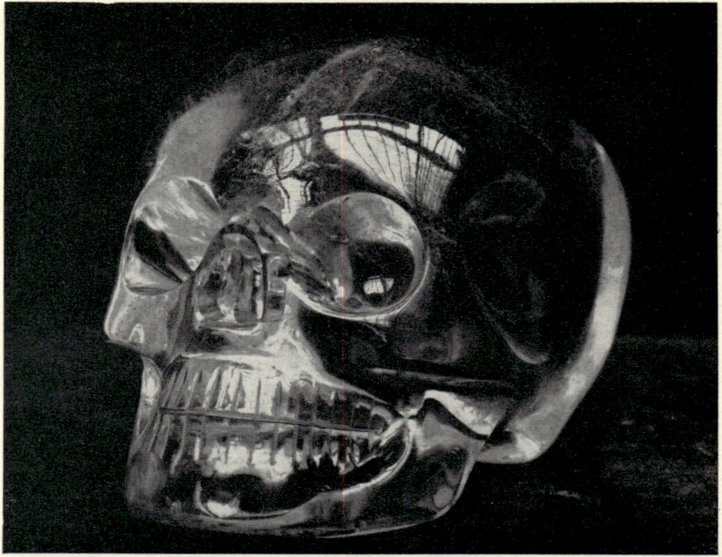

(*British Museum*)

15. MAYAN: *Chac-Mool Figure.* (*Museo Nacional, Mexico*)

16. MAYAN: *Maize Goddess*
(*British Museum*)

and organized priesthood developed together and the work of the sculptor served both alike as a symbol of supernatural and human authority. It was a necessary part of the great buildings in which authority was now centralized, the palace and the temple. On a large scale and in a permanent material it stood for a permanent order of things. The rulers of Assyria are portrayed in carved reliefs as invincible beings, triumphing in the hunt over the wild animals to which man was at last superior, or dominating a human enemy in size and strength. The Pharaohs of ancient Egypt are represented in huge statues as god-like beings.

At all times a purpose of sculpture was to give an image of the ideal, that is of something perfect of its kind. Even the cave-men have left female figures carved in mammoth ivory, not very beautiful to the modern eye, but presumably representing a type which answered to the artist's idea of the perfect woman. An image of power was the ideal of these early civilizations, however grotesque and fearsome it turned out to be. It was the great achievement of the ancient Greek civilization to conceive an ideal human figure, considered as a physical structure, perfect in the harmony and relation of all its parts. Free of the superstitious terrors which had oppressed earlier and less favourably placed peoples, the Greek mind gave a human stature and character to the gods it imagined. In sculpture, while reducing the gods to human size and aspect, the Greeks also raised the human being to a dignity and beauty which has profoundly affected the world's outlook.

An ideal of a related kind in the East is the

Buddha figure endowed with the superior qualities of a calm wisdom and detachment from material things. The purpose here again is to set an inspiring example. It is another aspect of purpose, apart from religious association, that sculpture has always served as a lasting record of exceptional individuals and their work and a reminder of historic events. In memorial form it has a continuity which ranges from the record of Roman victories carved on the Trajan column to the work of a modern artist which commemorates the bombing of Rotterdam in the Second World War.

In some ways the purposes may seem to have changed in modern times. Yet certain factors remain constant. There is still an element of magic in the process by which the ideas of the time take visible shape. Albert Einstein has said "The most beautiful thing we can experience is the mysterious. It is the source of all true art and science", and this remains true of sculpture in the twentieth century.

So far stress has been laid on the value of sculpture in informing us about various types of society and their ideas and beliefs. Yet, as a great form of art, it has a value distinct from this. In the service of religion the artist's role has been apart from that of the priest. Assuming, for instance, that the priesthood of the Aztecs of Mexico required certain features as the distinguishing mark of one of their weird deities, it was only the artist who could give dramatic power to the image. It would not exist without his imagination and skill.

The purpose of the sculptor as an artist is to give order and meaning to form. More than other artist she is a craftsman dealing with all the problems that arise from the handling and shaping of solid material. The beauty of his work is to be found not only in an idea but in his sense of what is appropriate to a given scale and of qualities inherent in the material he uses. Working on a monumental scale the master is concerned with those abstract properties of form, mass, rhythm, balance, which give an effect of grandeur. On a small scale and with varied materials to which it is suited, he may concentrate on exquisite detail and refinement of surface. Again, he will take into account not only the effect of solids but of sculpture as an enclosure or intersection of space, the latter being an aim which has particularly interested a number of sculptors in this century. Processes and materials are further described in the following chapter.

CHAPTER II

TYPES AND MATERIALS OF SCULPTURE

THE two main divisions of sculpture, carving and modelling, are in strong contrast. In carving a figure in the round it may be said that the artist already sees it in its completeness in the solid block before him and that his work consists in exposing it to view. This was the approach of Michelangelo as defined in one of his sonnets, rendered into English by John Addington Symonds as:

"The best of artists hath no thought to show
Which the rough stone in its superfluous shell
Doth not include: to break the marble spell
Is all the hand that serves the brain can do."

In the many statues that Michelangelo left roughed out or half-smoothly finished he might almost seem to have stopped at a certain point so as to show the figure dramatically breaking through its covering. Remarkable examples of this are the "Captives" in the Accademia, Florence.

The materials used by the carver also have their importance in his work. They vary from marble, stone of different degrees of hardness and texture to wood and ivory. Marble, finely textured, and lending itself to the production of a smoothly polished surface, may be called the classic material of sculpture. The Greeks used marble from Mount Pentelikon near Athens; and, from ancient Roman times, sculptors in Italy have drawn upon the marble quarries of Carrara near the east coast of the Gulf of Genoa, which provide black, yellow and green marble as well as white. Hard stones such as granite, basalt, diorite and hard limestone, local materials from the quarries of the cliffs flanking the desert plateau of Egypt, made for the simplification and boldness of treatment which is to be found in the monumental works of the Egyptian sculptor.

Wood, more easily worked, not only provides attractive variations of grain, of which sculptors have known how to take advantage, but allows of such elaborate craftsmanship as that of the wood-carvers of the medieval and Renaissance periods in

Germany and Austria. Ivory, which includes not only the tusks of elephants, but also those of the walrus and the teeth of larger animals, is especially suited to small, fine work. The size of the tusk necessarily limits the scale of the carving. The refinement obtainable can be appreciated in Indian, African and European examples—of the last of these, the Byzantine period has left many small masterpieces.

Modelling builds up an object from a soft material, instead of revealing that imagined as existing in its completeness in a solid block. Its textures and surface are not the inherent qualities of a hard material but the result of free manipulation. On a small scale it may be kneaded firmly into shape without requiring internal support. Models of clay and terracotta are hardened by firing in the same way as pottery. Indeed a great amount of work has been produced at various times which is half-way between sculpture and ceramics, calling into use those coloured glazes of surface which belong especially to the potter's art and craft.

Some may be left out of account in a study of sculpture; yet certainly under that heading we must include such beautiful things as the figurines in terracotta ("baked earth") of Tanagra and the Hellenistic period in Greece and the ceramic animals and figures of the T'ang dynasty of China. On a large scale, modelling is the first stage for the production of an object in more permanent material, bronze, the alloy of copper and tin, being traditionally used in casting. The large clay model is built on, and supported by, a metal skeleton, the armature. The highly technical process of hollow casting, known as "cire perdue" ("lost wax") has been practised for thousands of years. The figure is modelled in a layer of wax on a plaster core. The wax surface is covered with plaster and the whole fired, the wax melting and being emitted through channels cut for this purpose. Molten bronze is poured in to replace it and, when the plaster is removed after cooling, reproduces the wax layer in metal. The method thus roughly outlined was used by a sculptor of the prehistoric civilization of the Indus valley, *c.* 2500 B.C., in the vanished city of Mohenjo-Daro, to produce a statuette of a girl dancer, which was discovered in this century (Plate 3). Benvenuto Cellini, in his famous autobiography, has left an exciting description of how after various mishaps and trials he successfully used the process for his bronze masterpiece at Florence, "Perseus with the head of Medusa".

Metal has been employed in various other ways. Egyptian craftsmen made figures of sheets of bronze and copper beaten out on a core of bitumen. Craftsmen of the early Aegean civilization cast metal solid in a mould and chiselled it into the desired form. The elaborately sculptured salt-cellar made by Cellini for Francis I was worked in solid gold.

Just as there are a number of works related to pottery so also there is a half-way house between sculpture and the productions of the goldsmith and silversmith. It may be noted that the magic properties anciently and superstitiously attributed to sculpture attached in some degree to ornamental jewellery. A residue of the idea is to be found in the use of the word "charms" for objects of this kind. It may be noted also that the goldsmith's shop in Italy was the training-ground of many a great sculptor.

A material which has fascinated many modern sculptors is iron, and, in working iron, the artist employs, as distinct from carving or modelling, the technique of the blacksmith. The contemporary English sculptor, Reg Butler, for instance, gained practical experience for his work in metal in a village smithy. Many modern sculptors, or perhaps it would be more appropriate to call them "workers in three dimensions", have shown a preference for a variety of unconventional or new materials: wire, perspex and plastics of one sort and another. The practice of *collage*, the sticking together of pieces of material, borrowed from painting, has also been used in the production of a hybrid kind of sculpture built up of prefabricated materials, and sometimes partaking of the nature of both painting and relief.

Another main division of sculpture, apart from that between carving and modelling, is between sculpture in the round, standing free, either in the open air or inside a building, or independent of place; and that which has some close connection with architecture. The variations of architectural form throughout history have produced a corresponding variety in the sculpture designed as a part of it. It is only in modern times that the growth of a "functional" architecture and the development of the "curtain wall" of metal and glass has excluded sculpture from this close association with building.

The two arts are practically one in some ancient structures, as, for example, the Great Sphinx of Gizeh in Egypt, some 2,300 feet long and 66 feet high, carved out of a great mass of rock in the form of a lion with the head of the ruler, Chephren, and the temple of Abu Simbel, carved out of the solid rock, with its four colossal figures of Rameses II. The tomb-temples of Petra, the ruined city of Jordan, carved out of the rock face between *c.* 100 B.C. and A.D. 100, form another illustration.

Sculpture in the round, whether a primitive idol or so magnificent a figure of a presiding deity as must have been the colossal Athene by Phidias which stood in the Parthenon at Athens, was a necessary feature of the temple, but in many other forms it has been an actual part of the architectural fabric. The column, with its obvious likeness to the trunk of a tree, suggested the flowering of its capital into some natural shape. Thus the capitals of Egyptian columns were carved into the likeness of palm and lotus. A classical form of capital which has had a long tradition of use is that of the Corinthian column, adapted from the spiky leaves of the acanthus plant. The Gothic sculptors delighted in carving naturalistic plant forms.

The relief, the adornment of a surface, has been productive of many marvels of art. It has given scope to the artist for a free representation of incident or of a number of figures and objects in one composition. The extent to which the relief is "low" or "high", a shallow carving or a considerable projection from the surface, has depended on several factors. The low relief on the façade of an

Egyptian temple gave a sharp effect of outline in the brilliant sunlight, this graphic character being cultivated and completed by the addition of colour. Under other conditions of light, deeper modelling has sought to carry at a certain distance the stronger effect of light and shade.

Architecture sets problems and also gives the sculptor opportunities. The triangular form of the pediment raises the question of how to fill its narrow extremities effectively. There are many buildings with sculptured pediments in our cities where one can see how it has been tackled.

The church has traditionally made for the union of the arts with architecture. Perhaps no form of architectural construction has offered greater opportunity to the sculptor than the Gothic cathedral, with its external niches for statues, its recessed porch inviting adornment, its variety of interior structure and equipment suggesting the enrichment of detail. It is a special characteristic of the Gothic age that mason and carpenter were also artists whose imagination was at work in every part of a building, converting a water-spout into a fantastic gargoyle, carving a boss at the intersections of vaulting, a corbel where two arches meet, a finial at a bench-end, a grotesque or amusing detail on the misericord bracket on the underside of a choir-stall seat, adding the beauties of sculpture to font, pulpit and tomb.

The union is again to be found in splendour in the baroque buildings of the seventeenth and eighteenth centuries, when dramatic effects of perspective in wall and ceiling painting were closely linked with sculptured figures conceived in the same dramatic style.

A general survey of the sculptor's art and craft must include reference to the use of colour. We are used to seeing many statues of the past from which colour has disappeared and many of modern times in which it has never had a part. It might be thought that the sculptor was always content with the whiteness of marble or limestone, or the warm tone of bronze, modified only by the green patina which results from the oxidization of the metal. The natural colour of some materials used, green jade for instance, is such that an artist would refrain from what might be called "painting the lily". Yet in fact sculptors have never debarred themselves from polychrome effect in work in both wood and stone. Even the colossal Sphinx of Gizeh originally had a coat of painted plaster. The Egyptians lavished colour on their sculpture, the life-like appearance of the celebrated head of Queen Nefertiti being one striking instance among a number. Both classical and Gothic sculptors delighted to apply bright colours to their productions. The interior of a medieval church originally glowed with a brilliance of hue which we can now only imagine.

Various means have been employed. Sculptors have used special stones to give the appearance of nature to the white and pupil of the eye. The funerary masks of ancient Mexico were covered with mosaic patterns of colour. Both East and West provide examples of wood-carving elaborately painted and sometimes gilded. Wood lends itself to the process in three dimensions as well as in flat wood panels prepared with gesso for the work of the pictorial artist. Reliefs in enamelled brick were typical of ancient Babylon. Mention has been made earlier of the many forms of colour-glazed ceramic sculpture.

In the nineteenth century colour had so far fallen into disuse that the production of a "tinted Venus" (now in the Victoria and Albert Museum) by the Victorian sculptor, John Gibson, caused quite a sensation. Some perhaps may still prefer to contemplate sculptural purity of form without the distraction of colour, but of many great works it is an integral and necessary part. In recent times, when a mixture of media has become frequent, it has been freely used.

In spite of many changes in the course of its historical progress, sculpture has the continuity of an art in which craftsmanship is especially important. The chisel, the drill, the hammer, the file, which the modern artist uses in carving, are implements substantially unaltered since the age of Pericles. The modeller now, as in the past, relies principally on his hands. The illusions of space and form given by the painter are free of the sculptor's discipline, but the latter is a constant factor creating a close link between sculptors of every age and clime.

CHAPTER III

PREHISTORIC AND PRIMITIVE SCULPTURE

IT is a remarkable thing that all the forms of visual art later practised by the human race are already to be found in the Old Stone Age in the period which is speculatively dated between about 20,000 and 10,000 B.C. The caves of the Dordogne region of south-west France and of the French Pyrenees and northern Spain give examples not only of painting, drawing and engraving but of sculpture in its main aspects. The carved relief on a monumental scale is exemplified by the frieze of horses and other animals at Cap Blanc, a rock shelter above the River Beaune in the Dordogne. Nearly life-size, they were cut in bold relief with flint tools not much different from those of the modern sculptor. Traces of colour show that painting was already an auxiliary.

Modelling in clay is illustrated by the striking realism of two models of bison found in 1912 in the cave of Tuc d'Audoubert in the French Pyrenees (Plate 1). Male and female, each about two feet long, they are treated with a vigour and an appreciation of their character and bulk which make them comparable with the masterpieces of cave-painting such as have been found at Altamira in Spain and at Lascaux, near Montignac, in France.

A prehistoric form of sculptured ornament appears in the carving of the throwing-stick, a grooved rod into which a spear or dart could be fitted, giving greater momentum to the weapon when thrown. The figure of a mammoth carved in reindeer horn shapes one such throwing-stick from Bruniquel (Tarn-et-Garonne), now in the British Museum. Whether the carving signified ownership, or was credited with some magic property, we can only guess.

The human figure is more elaborately, though often crudely, treated in prehistoric sculpture than in painting. There are examples both in relief and in the round. The head of a girl, carved in ivory and simplified in a way that seems surprisingly modern, is a celebrated fragment of a statuette, found in 1892 and now in the Museum of Antiquities at St Germain-en-Laye (Plate 2). An equally famous relief in limestone is the "Venus of Laussel", a female figure holding a horn-shaped object, emphasis being laid on breasts and thighs as in other cave "Venuses". That such figures were connected with cults and ceremonies relating to fertility seems likely enough, though the nature and, indeed, existence of these cults is a matter of speculation alone.

The girl's head is assigned to the Aurignacian period, according to the classification which has divided the progress of prehistoric man into periods named after the places and regions of the cave discoveries. The Aurignacian (from Aurignac in south-west France) is the earliest of the cultures that have been distinguished. The "Venus of

Laussel" belongs to the Perigordian period which followed. Then came the Solutrean (after Solutré in Burgundy), apparently representing the inroad of a different people from the foregoing, who have left a good deal of sculpture. The last and greatest phase was the Magdalenian to which the Cap Blanc carvings belong. Rather apart from its wonders of animal painting, one may regard as sculptural rather than engraving some magnificent deeply incised outlines, such as those of a reindeer looking back (Musée de St Germain). We have to stretch our imagination to appreciate the immense span of time in which the art of the European caves was produced—far longer than the whole of recorded history. It was skill developed during thousands of years and its masterpieces represent a culture which can scarcely be called "primitive". Sculpture takes on a new aspect in later tribal communities, not yet organized in cities but having some measure of agriculture, possessing domesticated animals, and not relying solely on hunting and fishing. It is the age of the totem, the mask, the fearsome idol, though "age" is not perhaps quite the right word. It is not a question of steady development through the centuries, but of the art produced by a particular type of human society, which may be still primitive even if of quite recent date.

So we have to take a wider view of primitive sculpture in the world as a whole without respect to its date. Let us jump from the New Stone Age in Europe, when the caveman and his art disappeared some 12,000 years ago, to the coastal regions of north-east America, the south of Alaska and British Columbia, where weirdly carved totem-poles, masks and sculptured figures, representing gods and ancestral spirits, have been produced within the last two hundred years, some even in the twentieth century, by Eskimoes and Indians. The totem-pole, carved in redwood, was a kind of heraldry denoting the mythical animal origin of a clan. Carved and painted masks represented the supernatural beings with whom the sorcerer, or *shaman*, claimed to hold converse.

The nomadic course of the Indians southwards halts again in Arizona where the Hopi Indians, up to the present day, affirm their communication with the world of spirits by masks representing them, which are used in ritual dances.

In those numerous islands dotting the Pacific Ocean to the north and east of Australia, in the maritime areas of Melanesia, Micronesia and Polynesia, beliefs much the same in their general character find their expression among the native peoples. Funeral masks represented the spirits of ancestors considered to keep benevolent watch over the living. Other sculptured objects were assumed to become the temporary abode of gods during those periods of ceremony and dance when they were invoked. Images fashioned of wood, wickerwork, shell, painted with coloured earths, stuck about with fibres, feathers, hair and animals' teeth were as ferocious in aspect as the god of war and human sacrifice, Kukailimoku, of Hawaii, as grotesque as the masks of New Guinea and New Ireland, all showing the strange fantasy of form in which uncivilized man has expressed his idea of the mysteries of life and death.

The sculpture of a primitive society is not in itself necessarily primitive in the sense of being crude and unskilled. The best evidence of this is given by the art of the African Negro, especially as practised in West Africa and in the Congo. It was long customary for the white savant to regard all such works as products of savagery, of value only in the study of anthropology. That they have such a value is not to be denied, but it needed the artist and critic of art to point out, what the anthropologists in the age of African discovery had missed, that the Africans had a genius for sculpture.

Tribal organization produced many of those features which have been noted among the native islanders of the Pacific and the Indians of North America. The ancestral figure, the mask of mixed animal and human form, used in the ceremonies of religious cults and societies, the carvings which paid homage to deities and spirits, appear in Africa as in Polynesia, but in a greater and richer variety. The variety was due not only to the number of tribes, each possessing a community tradition of sculpture, but to differences of social structure and outlook, some of which have a definite explanation while others remain somewhat obscure. Nigeria gives the most striking examples.

It was only in the nineteen-forties that evidence of a prehistoric culture was found in the northern provinces of Nigeria in fragments of terracotta figures, dating, on geological and radio-carbon evidence, to the first millennium B.C. This *Nok* culture produced heads modelled with a powerful combination of realism and abstraction. It probably survived long enough into our era to influence the art of the Yorubas of Ife, which flourished in the thirteenth and fourteenth centuries A.D. The productions of Ife, however, stand alone in that extraordinary dignity and simplicity of life-like representation which has been called "classical" (Plate 40). It remains a question how a style which recalls that of the great age of Greece could have developed. All we can say is that from the practice of a cult which called for likenesses of dead chiefs there came a magnificent realistic art.

The Ife technique of bronze casting seems to have been transferred to neighbouring Benin, famous for the bronzes and ivories of the "court style" which reached its highest pitch of development in the fifteenth and sixteenth centuries. Unlike other African sculpture it was a city art, with a political rather than religious background, conveying the power and wealth of the royal *Oba* who ruled in a great mud palace. There are many examples of the bronzes and ivories of Benin, brought back after the British expedition of 1897, in the British Museum, which show the high degree of technical skill and realistic representation achieved by the artists of this African state (Plates 18, 41). We see the *Oba* in majesty, with his sceptre, every detail of his costume and adornment carefully and decoratively detailed, his attendants on either side in a bronze of masterly execution and finish. Among the ivory masks found in the palace of the deposed *Oba*, Ovonramwe, the British Museum has a superb head, dating evidently from the time of the arrival of the Portuguese in the late fifteenth and sixteenth centuries, the tiara which offsets the simplicity of feature being composed of an alternation of Portuguese heads and emblems of the sacred mudfish.

Benin sculpture has an historical development which can be traced for some four hundred years until its decay in the nineteenth century. It seems to have grown formalized in the course of time, becoming increasingly ornamental in detail during the eighteenth century. In this evolution it is different from the tribal sculpture of religious purpose, which is much more austere and simple and without so distinct an historical background. A number of different tribes produced local variations in the form of mask and fetish figure; but if one seeks for a general term by which to describe the African genius in sculpture it may be found in the decision and force of wood-carving. Like Michelangelo, the African carvers in their way had a sense of the complete figure, inherent in, and waiting to be released from, the block. The Baluba art of the Congo provides some extraordinary examples. The bold simplification of the mask and the emphasis with which features were cut out in decisive planes had a lesson to teach to the European artist in the twentieth century.

Where African sculpture is inimitable is in the complete fusion of idea and purpose with art and craft. To appreciate it fully, one has to have something of both the anthropologist's and the art critic's point of view. The European artist could not produce anything like the same result for the simple reason that he was remote from the tribal way of life. Yet artist speaks to artist in spite of every barrier of date, place or belief. They can borrow, without imitation, for a purely artistic purpose. The sharply cut planes of the African mask contributed to the new idea of structure which was evolved by painters in Paris and which launched the Cubist movement. The elongation of an African mask inspired Modigliani to produce a head in which the same device was strikingly employed, though the difference of race and outlook produced an entirely different effect. A number of modern artists have been influenced, not so much by any desire to produce similar results as for a reference, which they have found inspiring, to basic principles of the art. Sir Jacob Epstein formed a great collection of primitive and ancient works of sculpture, but his own work derived little or nothing from them, unless it was in the incitement they gave to simple and forceful expression. His affection for sculpture that came from the Gaboon, the Sudan, Ghana and the Ivory Coast, Nigeria and Oceania, as well as from the civilizations of the distant past was that of the fellow-craftsman. The lay person may be reminded by it that African and other forms of folk-sculpture in primitive communities are not necessarily primitive as art but may show a real mastery.

CHAPTER IV

THE SCULPTURE OF ANCIENT EMPIRES

MESOPOTAMIA

The "folk" or "primitive" sculpture described in the previous chapter is "outside history" in being the product of a type of primitive society which might exist at any period and indeed still exists today in regions such as New Guinea, for example. The sculpture of ancient cities and the empires of which they were centres have a more clearly defined historical existence, taking us back to epochs long before the Christian era, when man began to live an organized, urban life, and coming to an end with the civilizations that supported them.

The region known as Mesopotamia, that is to say, the plain enclosed by the rivers Tigris and Euphrates and also the adjacent regions on either side, gives a series of great early examples. In succession there flourished city-states, warlike of necessity, for the absence of any natural frontiers incited various phases of conquest by the different surrounding peoples, but with a highly developed architecture, legal system, export trade in grain and a system of writing (made necessary by legal and mercantile transactions), developing in some 3,000 years from *c.* 3500 to 500 B.C. Sculpture, represent-

Map of Ancient Greece showing places of historical interest

ing authority both supernatural and human, reflects the character of these city cultures and was now closely associated with architecture.

The earliest centres of art were the cities comprising the kingdom of the Sumerians, in southern Mesopotamia, not far from the Persian Gulf (see map). One of the great triumphs of modern archaeology, the excavations of Sir Leonard Woolley at Ur, revealed the early splendour and refinement of sculptural decoration in a city already many centuries old before 3500 B.C. The art of Ur is notable mainly for its use of precious metals and rich detail rather than for its monumental character. Masterpieces of sculpture are the small bull's head in gold, originally the ornament of a harp or lyre, and the statue of a goat, in gold, lapis lazuli and shell, rearing before a golden tree to which its front legs are bound by silver chains, both now in the British Muscum (Plate 21). Half sculptural and half-pictorial, is the famous so-called "Standard of Ur", a wooden structure decorated with scenes in coloured mosaic representing a royal banquet and related processions of servants and cattle. It dates from the third millennium B.C.

An impressive form of monumental sculpture comes from Lagash, a neighbouring city to Ur after the Semitic conquest of Sumeria, portraying the priest-king Gudea who governed the city by "divine right". Several versions exist (British Museum and Louvre) of this free-standing statue, symbol of authority, carved from diorite the most permanent of hard stones. The material made for the simplicity of treatment which creates the impression of both power and repose. These images date from the second millennium B.C.

This serene and monumental form of sculpture is typical of the period when the Elamites, whose capital was Babylon, ruled over Sumeria. The oldest of all forms of script, the wedge-shaped cuneiform writing, is combined with sculpture in the basalt cone, now in the Louvre, recording the legal code of the Babylonian ruler, Hammurabi. The bas-relief in which Hammurabi is shown in prayer before the sun-god Shamash, who dictates the law, indicates once again the divine support invoked for human institutions and the visible testimony to this effect that sculpture could give.

The most remarkable forms of sculpture in Mesopotamia were those of the warrior race of Assyrians of the northern region who dominated the whole of western Asia from the ninth to the seventh centuries B.C. and built great palaces at Nimrud and Nineveh. Reliefs and colossal figures were designed to convey the invincible force bestowed on the Assyrian rulers by the national god, Ashur. The combination of human characteristics with the attributes of powerful animals, already noted in primitive sculpture, appears in the Assyrians' huge winged and man-headed lions and bulls, which stood at the palace doorways as guardian spirits. The triumph of man over the animal world and of the Assyrian over all his enemies was celebrated in the great reliefs which are masterpieces of ancient art.

The spirit of this art is summed up in the lion hunt of Ashur-bani-pal, the Assyrian king of the seventh century B.C. (British Museum) (Plate 12). Impassive, formally bearded, the king in his chariot and his attendants with spears and bows mete out destruction to lions and lionesses, the character and movement of which are portrayed with wonderful understanding and skill. This is one of many scenes of the chase and war which bring an ancient race vividly to life. The palace of Ashur-bani-pal covered some six and a half acres of ground, its throne-room measured 45 by 15 metres. The reliefs of the walls, now without colour, were brilliantly painted. That in the women's quarters and elsewhere the artist was not confined to military propaganda has been shown by excavations later than those of Sir Henry Layard. The superb ivory head now known as the "Mona Lisa of Nimrud" was found at the bottom of a well on the site in 1952. Now in the Iraq Museum at Baghdad, it dates from the reign of Sargon, *c.* 715 B.C. and its "archaic smile" has suggested that it belongs to a type of Asiatic sculpture which inspired or provided a model for early Greek sculpture. Another great find of the same date was a beautifully executed carving of a lion killing a Negro in a field of lotuses, though this was of Phoenician origin and showed a sensitiveness of a different order from that of the scenes of the royal hunt.

Trade and conquest over a long period caused a constant interaction between the peoples of Western Asia. The Hittites of Asia Minor, or Anatolia, also had a flourishing empire between 1600 and 1200 B.C., were closely associated in both respects with the Assyrians, and their sculpture is related in character to that of Mesopotamia. The Mesopotamian influence extended to the trading capital of the Phoenicians on the Mediterranean, the great city of Tyre. Assyrian and Babylonian supremacy was finally destroyed by the invasion of the Persians from the east, Babylon falling to Cyrus, the Persian king, in 509 B.C. Under the Persian dynasty of the Achaemenids, at Susa and Persepolis, where Cyrus had a great palace, sculptors inherited the style of relief and the bull and lion forms of Assyrian art, though these took on a more decorative and conventional character. The famous frieze of bowmen of the imperial guard (called "Immortals" by the Greeks), from Susa, like the one from Persepolis (Plate 13), is a coloured relief in enamelled brick, now in the Louvre, illustrating the rich but decadent art of the period between the seventh and fourth centuries B.C.

Archaeological research has shown that art developed in Western Asia earlier than in Egypt and was an influence on the art of the Nile, though for centuries they ran a parallel course, Egypt attaining a separate creative height which still amazes us today.

EGYPT

The sculpture of ancient Egypt is one of the most complete and magnificent memorials of a way of life and mode of thought that any country has left. Based essentially on the idea of a life continuing unaltered after death, it gives for that very reason a complete picture of the Egyptians' actual existence, their features, dress, ceremonies, festivities, adornment and everyday occupation. Representing the idea of permanence, it is an art which conforms to unchanging standards for a period much longer than the whole of our Christian era.

The long and stable tradition of Egyptian art was due to several factors. The country was made one at an early date, its unification being attributed to the legendary King Menes in the third millennium B.C. Wars and invasions had little effect on its internal life until Greek and Roman conquest, a little before the beginning of the Christian era, destroyed its native culture. The settled rule of king and priesthood gave sculptors and other artists security and dictated the nature of their work. Families of craftsmen worked in the local quarries, handing on their skill through the generations and the centuries.

Geography played its part. Along the fertile valley of the Nile, populous and wealthy cities flourished, from Karnak Luxor and Thebes in Upper Egypt to Memphis in Lower Egypt, near the site of the present Cairo. The ridges of the desert plateau provided the granite and limestone for their monuments. Sculpture gives a view of a wealthy and intelligent society, not so sombre in religion as to be unable to enjoy music and the dance and to apply artistic skill to all sorts of luxuriously designed objects of use. The forms of sculpture are various. In the Neolithic period a skill in carving already appears in vases and other vessels hewn out of stone such as have been found at Sakkara. A form of early relief appears in palettes or memorial tablets, apparently influenced in style by contact with Sumeria. Low relief, giving the effect of outline, used on both the exterior and interior walls of temple and tomb, developed with a constant building activity into a main feature of art. There is a rigid convention in the figures carved in profile and a characteristic stance with one foot in advance of the other.

The reliefs however which are numerous at every period are full of vigour and variety of interest. The Egyptian sculptor was able to combine a great deal of lively observation with the formal rules of his craft. In the reliefs which celebrate military triumph, the racial character of the prisoners is sharply distinguished. Movement is rendered with wonderful verve in the high kicks or twisting contortions of the dance. The harvesters, the flower-gatherers, the bird-catchers among the papyrus plants, the servants with their panniers of bread and fruits are seen as they were in reality. From the representations of musicians we can gain an exact idea of the make-up of an Egyptian orchestra—flute, harp, oboe, clarinet, trumpet, drum.

Portrait sculpture produced many masterpieces which range from a serene and abstract aloofness conveying the majesty of priest and king to an intimate humanity. The magnificent statue in diorite of King Chephren, builder of the Second Pyramid at Gizeh (Cairo Museum) is an example of the former. In the less severe material of wood is the celebrated statue of Ka-Aper, also carved more than 2,000 years before Christ, so realistic in its expression of kindliness and informal pose that when it was discovered one of the Arab workmen remarked that it was exactly like the local sheikh. The priest and high official, Ka-Aper, thus gained the familiar title of "Sheikh El-Beled" (Cairo Museum) (Plate 6). The importance of writing and written records gave the scribe an eminence which appears in sculpture. Of the representations of the seated scribe with a roll of papyrus before him there is a famous example from Sakkara in painted limestone (Louvre, *c.* 2500 B.C.), the features conveying an acute intelligence (Plate 7). A combination of life-like features and rigid pose is to be found in the painted limestone figures of Prince Rahotep, priest and military commander, and his wife Nefert ("the beautiful one"). Realism in which the artist was entirely free belongs to the one artistic revolution of ancient Egypt, that of the reign of Akhnaton, the unconventional ruler who established a new religion, the single worship of the sun, and encouraged a new art.

His own inbred and decadent features appear in some remarkable portraits of *c.* 1360 B.C. The beauty of his Queen, Nefertiti, is immortalized in the head in the Berlin Museum and (even more impressive as a work of art) the unfinished quartzite head in the Cairo Museum (Plate 8). At all times in Egyptian history, landscape as well as ideas of religion and state had prompted colossal conceptions of which the Sphinx at Gizeh is an early example. The father of Akhnaton, Amenophis III, had returned to this idea in the so-called "Colossi of Memnon" at Thebes which gave giant stature to his image. The art encouraged by Akhnaton seems a protest against this kind of vain glory. The return to the image of power after him marks the end of his influence. A more conventional form of art, astonishingly luxurious as it is, appears in the tomb of Akhnaton's successor, Tutankhamen, that famous tomb found almost intact in 1922 (unlike most of the royal tombs which were pillaged even in ancient times). The return to imperialism is signalized by the later dynasty of the Rameses. The colossus once more appears in the statues of Rameses II at the rock temple of Abu Simbel.

Yet some of the most delightful of Egyptian plastic works are small scale models in three dimensions which, like the reliefs of secular scenes, bring Egyptian life vividly before us. They projected its various aspects into the supposed future after death, the spinners and weavers at work, the woman servant brewing beer, the company of soldiers with shields and lances, the Nile boats with their graceful lines, the fishermen hauling in their catch. These models from the tombs are executed with a wonderful realism.

Archaeology has established thirty-six dynasties and a series of broad historical divisions in a period which extends over some 2,500 years before the birth of Christ. Except, however, in Akhnaton's reign, art remains amazingly consistent in character during this long time. It existed unchanged in essentials just as long as the society and the beliefs which brought it into being. It disappeared with the alien dominance of Alexander the Great and the Ptolemies between 332 and 30 B.C., and of Rome between 30 B.C. and A.D. 395. The waves of conquest caused a weakening of old ways of thought which is first reflected in a mechanical copying of earlier sculpture and then in a mixture of very different styles as Greek, Roman and Egyptian ideas came into collision or attempted union. The spread of Christianity in the second century A.D. brought sculpture and all the other arts that had flourished under the Pharaohs to the point of no return.

ANCIENT AMERICA

Like the sculpture of Egypt, that of central America, in the period before Columbus, flourished in populous cities dominated by a religious system. A superficial likeness between the pyramid constructions of both lands at one time led to fanciful theories of contact between them, though beyond the fact that they represent a similar stage of human progress they are as far apart in character as in physical distance. Maya and Aztec art gives form to gods and rituals more cruel, and ideas more pessimistic, than those of Egypt. There is a strong contrast also in artistic character. Whereas in Egyptian sculpture admiration goes to its serenity, its delicacies of line and profile and the interest of its subject-matter as a realistic record, Maya and Aztec sculpture is full of a disturbing energy, a concentration of force which has more deeply impressed and influenced the sculptors of the present day than any other work of ancient times.

The origins and early stages of art in Central America are still imperfectly known. At some period the wandering tribes, moving down from the north, settled in various regions to fixed abodes and an agricultural life. In consequence, while there is a

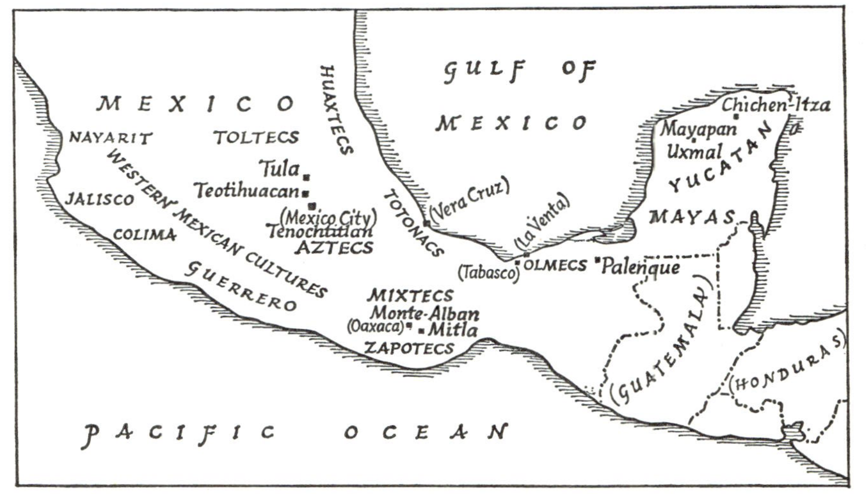

Centres of Pre-Columbian, Mexican and Maya Sculpture

certain family, American-Indian, likeness in what they produced, there are also various regional differences, as well as differences of date in artistic production. The earliest civilizations were those of Western Mexico and the so-called "Olmec" culture on the Gulf of Mexico, going back to more than a thousand years B.C. A classical period between *c.* A.D. 300 and A.D. 900 comprises the cultures of the central Valley of Mexico and that of the old Maya empire, which extended over the modern states of Yucatan and Campeche, the present Republic of Guatemala and British Honduras. The Toltecs and the Aztecs bring us to the period between the tenth and sixteenth centuries A.D. The development of art was complicated by the rise and fall of cities and the movement of peoples, the Toltecs for instance migrating southwards and imposing themselves on the Mayas in the period of the "New Empire". In general, however, there is an evolution of city from village and the emergence of an art entirely dedicated to religion, taking its most bloodthirsty form with the warlike Aztecs, whose empire it was the main object of the Spanish to destroy in the conquest of 1521.

The Olmecs are a riddle still, though their culture seems to have lasted for more than a thousand years in the first millennium B.C. and the first millennium A.D. All that remains of it now are colossal heads and altars carved in stone, and small sculptures and ornaments in jade. In the garden museum of La Venta, the visitor can appreciate the grandeur of the carved heads, priests and nobles perhaps, yet of a strange type incorporating char-

17. AZTEC: *Mosaic Mask*
(*British Museum*)

18. AFRICAN: *Benin Bronze, Oba and his Attendants*
(*British Museum*)

19. INDIAN: *Bronze Buddha, Sultanganj, Bengal. Seventh century*
(*Birmingham Art Gallery and Museum*)

20. MINOAN: *Capture of the Bull, Vaphio Cup. Sixteenth century B.C.*
(*National Museum, Athens*)

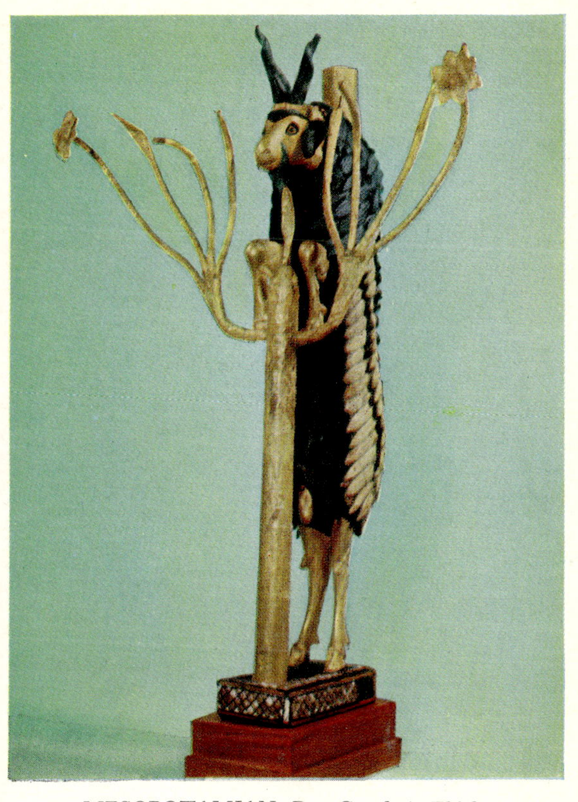

21. MESOPOTAMIAN: *Ram Caught in Thicket, Ur*, c. *3500 B.C.*
(*British Museum*)

22. GREEK: *Archaic Figure, the Peplos Kore,* c. 530 B.C.

(*Acropolis Museum, Athens*)

23. GREEK: *Charioteer of Delphi. Bronze,*
c. *470 B.C.*
(*Delphi Museum*)

24. GREEK: "*Ludovisi Triptych*", *showing Aphrodite rising from the Sea. Fifth century* B.C.

25. GREEK: *Horsemen, the Parthenon Frieze. Fifth century* B.C.
(*British Museum*)

26. GREEK: *The Discus-Thrower* (*after Myron*)
Bronze, c. 450 *B.C.*
(*National Museum, Rome*)

27. GREEK: *The Laocoön. Hellenistic group by sculptors of Rhodes. Second or First century B.C.*
(*Vatican Museum*)

28. GREEK: *The Victory of Samothrace. Rhodian sculptor, third or second century B.C.*
(*Louvre*)

29. ROMAN: *Detail from the Trajan Column. First century A.D. (Army crossing the Danube on a bridge of boats)*

30. ROMANO-ETRUSCAN: *The She-wolf*

(*Museo dei Conservatori, Rome*)

(left) *Nero*
(Capitoline Museum, Rome)

(right) *Julia, Daughter of Titus*
(Naples Museum)

31. ROMAN PORTRAITURE.

32. BYZANTINE:
"Harbaville Triptych." Ivory. Tenth century

(Louvre)

acteristics of their jaguar god, especially in the feline nature of the mouth. In the humid heat of this tropic lowland the effort that transported these huge monoliths from a far distance is the more a matter of wonder. What invasion overturned the temples, leaving only the massive debris of sculpture, is not known, though the jungle growth of vegetation has probably secrets yet to reveal to the archaeologist's research.

In the urban centres of Mexico and Central America, dominated by the great stone pyramids, the steps of which led up to the temple at the summit, sculpture flourished with architecture. Sculpture, with some exceptions, as on the west coast where art was comparatively free and secular, was closely bound up with the conception of sombre and ferocious deities, of cruel ritual. A famous example of classical Mayan art (British Museum) is the relief (A.D. 726) which shows a priest with plumed staff, before a devotee who tears his tongue with a cord bristling with thorns. The gods—the fiery serpent symbol of the sun, Coatlicue, the earth mother, and Tlaloc, the rain god—seem full of vengeful power. The image of death shadowed this Indian culture, especially that of the warlike Aztecs who tore out human hearts, the symbol of life, to sacrifice to their gods. It is likely that they employed sculptors from the more artistic peoples whom they conquered, though from the thirteenth to the early sixteenth century Aztec art becomes distinct. The skull carved out of rock crystal and the actual skull converted into a mosaic mask of turquoise and obsidian in the British Museum are among the most remarkable remains of this morbid culture (Plate 14).

Even so, morbid as it was, a distinction is to be made between the cruel priestly cult and the genius of the sculptor who could create not only terrifying images but forms of great vigour and beauty. The greatest artists of this pre-Colombian America were the Mayas, who excelled in architecture and sculpture, and the Toltecs, who had their own capital in Central Mexico at Tula until A.D. 1168 when they moved southwards, their history becoming intertwined with that of the Mayas. An impressive product of the Toltec-Maya culture is the type of reclining figure, "Chac-Mool" (Rain Spirit), to be seen in the wonderful remains of the Maya city of Chichén-Itzá in Yucatan (Plate 15). The Chac-Mool figure at the entrance of the Temple of the Warriors at Chichén-Itzá has a grandeur akin to that of the carvings of Henry Moore in our time. There are funerary masks of a classic simplicity and dignity, like those of the city of Teotihuacan which flourished from A.D. 300 to 900. The ceramic sculpture of the Zapotecs of Monte Alban and the figured ornaments of the Mixtec goldsmiths (of the present state of Oaxaca) indicate the range of plastic art before the Spanish conquerors arrived.

The other great civilization of America before the European advent, that of Peru and the Incas, which came to maturity at the same time as that of the Aztecs, offers no similar works of sculpture on a grand scale. Works in gold and silver, metals which for the Indian did not possess the economic value attached to them elsewhere and since, were melted down by the European conquerors and little remains on which to form an estimate of sculpture in these materials. The most individual form of Peruvian art is the vase-portrait, the vessel of terracotta modelled in the shape of a human head. The great centres of production were the Moche and Chimaca valleys of the northern coast, the flourishing period, sometimes called the "Master Craftsman" period, lying between the beginning of the Christian era and *c.* A.D. 1000. It was, that is to say, long before the establishment of the Inca Empire from Ecuador to Chile, in the two hundred years preceding the Spanish conquest. Probably the portraits of nobles, the pottery heads attain a remarkable degree of realism.

The study of ancient American sculpture, extending from the Mexican highlands above Mexico City to the tropical lowlands of Yucatan, where a great number of sites still remain to be explored, and into central and south America, is that of a rich and complex art with a history, in all, of some two thousand years. There is no region of the world perhaps in which even now there remain so many possibilities of fresh discovery.

CHAPTER V

THE MEDITERRANEAN GENIUS

THE Mediterranean sea, from ancient times a meeting point of east and west, was a link between lands as much in art as in trade. As a trading centre, the island of Crete became prosperous at the same time as Egypt with which it had close relations. It seems likely that its first rulers came from Asia Minor. It became the centre of an Aegean civilization with cities also on the Greek mainland. Its culture developed between the third millennium B.C. and 1400 B.C. when Crete fell to invasion by the northern people known as Dorians. Sculpture, though on a small scale, is one of its achievements in art.

One has the impression of a luxury-loving, athletically minded race, less conditioned than the Egyptians by the weight of a religious system or the monumental architecture and sculpture which accompanied it. The Cretan deity was an earth goddess, a mother figure. Figurines portray priestesses and ceremonial dancers, with tight-waisted, flounced skirts curiously suggestive of the Victorian crinoline. The legend of the Minotaur suggests some remote phase of prehistory when the Cretan priest-king, perhaps made fearsome by animal head-dress, half-man half-bull, received tribute and sacrifice, though surviving representations

Ancient Centres of Western Asia and Egypt

of the bull are naturalistic and have a different reference.

The bull, like the lion, was in sculpture often a symbol of power, but Cretan representations of the animal are not so much symbolic as suggestive of the bull-fight or some contest of a sporting kind, which is also depicted on the Cretan frescoes. A beautiful gold cup sculptured in relief (National Museum, Athens), dating from the sixteenth century B.C., shows the capture of a wild bull, in a superb but entirely naturalistic design (Plate 20). The earliest monumental sculpture of the Greek world is the famous Lion Gate at Mycenae on the Peloponnesian mainland, *c.* fourteenth century B.C. Its two lions facing inwards towards the column on which their forepaws rest, form a motif borrowed from Mesopotamian sculpture. The lively secular outlook of the Cretan artists found principal expression in the art of the goldsmith and carvings on a small scale.

In some ways the lively observation they showed is akin to that of the Greek art which followed, though the Greeks were racially different (this may be said without exact knowledge of the racial components of ancient Crete and its outposts). The invaders of the first millennium B.C. who founded the civilization we now call Greek (whatever their racial character) may nevertheless be said to begin again in art. The end of Cretan civilization was the prelude to a more splendid evolution. A form of art much more primitive than that of Crete and Mycenae is the starting-point. Between the tenth and eighth centuries B.C., the "Homeric" period, small statuettes of bronze and ivory with a rough simplicity of form were produced as votive offerings. Shields were as elaborately ornamented as that of Achilles, as described by Homer, with motives derived from Mesopotamian art. In the seventh and sixth centuries B.C., the so-called "archaic" style developed. Large statues in the round, carved in marble, were placed in stone temples and sanctuaries as offerings to the gods or representations of the gods themselves.

Contact with Egypt brought with it the influence of Egyptian sculpture, which appears, in the earlier examples, in a rigid symmetry and conventional stance, the left foot stiffly advanced before the right, the features set in that fixed expression known as the "archaic smile". From the sixth to the fifth century B.C. a rapid and remarkable development can be seen in the characteristic figures of the nude male and the clothed female. The practice of athletic sports and the Greek spirit of systematic inquiry seem both to have made for an understanding of the human form which freed its rendering in sculpture from the conventions borrowed from the Orient. The fixed "smile" becomes a more natural expression of serenity. Natural movement is studied, as in the pose of a kneeling archer, the Hercules of Aegina, early fifth century B.C. Drapery becomes a beautiful system of rhythms. A fearless intelligence, free of the inhibitions of more primitive races, conceives the gods as human beings and human beings as capable of a godlike perfection.

From about 480 B.C. there is a succession of masterpieces and at the same time, even if we know their work only by later copies, we become aware of the existence of individual artists. Sculpture had at last ceased to be anonymous; it was a personal exercise as well as a function of state and religion. The sculptor, Myron, can be appreciated, from Roman copies, as one with a complete grasp of the flow of athletic movement, in the famous "Discus-thrower" (Plate 26), *c.* 450 B.C. Polykleitos was an artist who established a canon of human proportion as in his "Doryphoros" (athlete with a javelin), *c.* 450–40 B.C.

The last stage of the archaic period is represented by such magnificent works as the "Charioteer of Delphi", *c.* 478 B.C. (Plate 23), and the Zeus in the pose of hurling his thunderbolt, *c.* 460 B.C., found in the sea off Cape Artemision and now in the National Museum at Athens. They might equally well mark the beginning of the "classic" period. The beauty of composition in relief is superbly represented by the "Ludovisi triptych" (*c.* 470 B.C., Thermae Museum, Rome), part of an altar which seems to show Aphrodite rising from the sea (Plate 24). The later half of the fifth century shows the full development of classical sculpture at Athens. Sculpture, it may be noted, had not been confined to that city. Great earlier centres were at Olympia (the temple of Zeus), at Delphi, the sacred abode of the oracle on the slopes of Mount Parnassus, and at the island temple of Aegina in the Gulf of Athens. The building of the Parthenon at Athens concentrates Greek effort in sculpture between 447 and 433 B.C.

The presiding genius was Phidias, one of the most revered names in the history of sculpture, though it is not now possible to assign to him any specific work. He designed the great statue of Athena for the temple but only a faint inkling of its grandeur is given by a remaining small copy. The sculptures of the Parthenon vary in conception and technique, though this was in part due to what was appropriate to pediment and frieze, internal and external. Yet every part was touched by that inspired spirit on which Plutarch remarked, in his *Life of Pericles*, the "unaging soul mingled in their composition".

The themes of the sculptors were both legendary and contemporary. The magnificent fragments of the west and east pediments that can be seen in the British Museum are from compositions representing the rivalry of Athena and the sea god Poseidon and the birth of Athena. The reclining male figures, the Dionysus and Ilissus, the female figures (the folds of whose dress are so expressively carved), the horse's head of the moon-goddess Selene's chariot are each masterpieces of sculptural form. The boldly carved exterior tablets (metopes) between the vertically grooved triglyphs of the exterior represent legendary battles, of gods and giants, Greeks and Amazons, Lapiths (mountain folk of northern Greece) and centaurs. In contrast is the great frieze running round the inner colonnade representing the Panathenaic festival, the yearly Athenian holiday. Its procession of horsemen, chariots, elders, musicians, vessel-bearers, sacrificial animals, maidens and magistrates, in escort of the sacred garment with which the statue of Athena, was to be draped, is akin to some masterpiece of musical composition, consistent in its main theme but passing through many and complex variations. In a feeling for actuality raised to a higher artistic power by rhythmic movement, the Parthenon frieze is an incomparable work (Plate 25).

Later classical sculpture shows two main directions. A calm perfection of feature and grace of form distinguishes the fourth century sculptors, among whom Praxiteles stands out. The marble statue by him of Hermes with the infant Dionysus (Olympia) *c.* 350 B.C., sums up the qualities of the ideal then aimed at. Yet there is a different tendency towards restless movement and emotional expression. The unrest can be seen in the frieze from the Temple of Apollo at Phigalia, *c.* 420 B.C. (British Museum). The theory has been put forward that it reflects the stormy emotions of the Peloponnesian War between Athens and Sparta (431–404 B.C.). Yet it seems also an expansion of artistic aims which is represented by the work of Skopas and Lysippos. This involved both vigour of action and unconventional pose.

The trend provides a separate name for the sculpture of the third century onwards. The Greeks had always called themselves "Hellenes" and their country "Hellas", but in modern usage "Hellenistic" describes this later period alone. It comprises the work done in local schools such as flourished in the island of Rhodes and at Pergamon in Asia Minor. The technical achievement of the Rhodian school had a dramatic evidence in the colossal figure of the sun-god, one hundred feet high, which once bestrode the harbour of Rhodes. One of the wonders of the ancient world, it was destroyed by earthquake after a half-century. An extant wonder is the "Victory of Samothrace" (Louvre) (Plate 28), *c.* 190 B.C., the figure alighting with a wonderful flutter of draperies on the prow of a galley. A final masterpiece of emotional complexity was the "Laocoon", the work of three Rhodian sculptors showing the priest of Apollo and his two sons, pursued by the wrath of the gods, in the toils of serpents (Vatican, first century B.C.) (Plate 27). The "Venus de Milo" is another work, once considered a supreme triumph of classical art, belonging to the Hellenistic second century, and an academic adaptation of earlier styles.

Greek sculpture, throughout its evolution, shows a sustained power which does not attach a necessary difference of quality to its archaic, classical and Hellenistic forms. At one time, in our own century, there was a tendency of criticism to exalt the primitive vigour of early work at the expense of the classical period. At an earlier time Hellenistic work was considered "decadent", the fifth century alone being considered the great period. Each period in fact has its inspiring works. Yet the wonderful evolution had probably run its course by the first century when Greece became subject to Roman rule. Greek masterpieces were transported in quantity to Rome, and the demand for copies was so great that craftsmanship was largely diverted to their production. Surviving copies, inevitably not as good as the originals, have at least the merit of preserving an idea of their nature.

The influence of archaic Greek sculpture is to be

seen in that of the Etruscans, a race settled in Italy between the Arno and the Tiber, coming, it is supposed, from Asia Minor before 800 B.C. and attaining a flourishing empire in the seventh and sixth centuries B.C., long before Rome became a centre of power. Their art is a link between that of Greece and Rome. The presence of Greek colonists in Italy and trading connections by sea with Greece itself made for a contact of cultures. The Etruscans portrayed the gods of Greece, as for instance in the remarkable Apollo, part of a mythological group from the Temple of Apollo at Veii, *c.* 500 B.C. (Villa Giulia, Rome). The famous "archaic smile" appears in this work, attributed to a sculptor called Vulca.

A difference from Greece appears, however, in a more sombre view of religion, evident in the demonic shape given to creatures of the Underworld. Another difference is to be found in the Etruscan cultivation of portraiture, as in the funerary sculpture of tombs and in portrait statues. They were different in their exact realism from the idealized portraits of poets and philosophers in Greek sculpture.

When Etruria came under the dominance of Rome in the third century B.C. there was a fusion of the two cultures. Etruscan sculptors worked in Rome. The superb bronze she-wolf of *c.* 500 B.C. in the Capitol Museum, Rome (Plate 30), is an application of Etruscan workmanship to the symbol of Rome's legendary foundation. Both peoples had something to contribute to the sculpture in which Rome became distinct. Portrait sculptures were made with a harsh fidelity to the features of individuals (Plate 31). Other works were produced as records of campaigns and victories. It was, according to the exhortation which Virgil puts into the mouth of Anchises in the *Aeneid*, the part of the Romans to rule rather than to hammer beautiful forms from bronze. While the Romans of the imperial age delighted as patrons in such replicas of Greek masterpieces as the "Medici Venus" and the "Apollo Belvedere", their own art was entirely uninfluenced by the idealism attached to the figures of gods and goddesses or the idea of an art produced for its own sake.

Instead we have a gallery of intensely living personalities, the austere features of Republican heroes, Caesars strong, kindly, weak or depraved; their consorts also; historical documents of the first order. The Greek intimacy with the gods disappears, the beliefs of a variety of conquered races are indifferently absorbed. A characteristic form of Roman sculpture is the relief which winds round a triumphal column, to be "read" in detail as a series of episodes rather than to be viewed, as the Parthenon frieze may be viewed, as a symphony of form. On the hundred feet high Trajan column at Rome (Plate 29), erected A.D. 113–14 to commemorate the victorious campaigns of Trajan in Dacia, one can observe the crossing of the Danube, the building of a fortification, the varieties of military costume. The column and the triumphal arch are the vehicles of this descriptive sculpture. It had no place, however, in the great functional products of Roman architectural engineering. Yet, with all its short-comings in respect of beauty, Roman sculpture, with its many reflections in the provincial schools of the territories the Empire dominated, contained the seeds of future development.

CHAPTER VI

SCULPTURE IN THE FAR EAST

In the general pattern of the ideas which it represents the sculpture of the Far East has several parallels with that of the West. There is the attempt to give shape to the forces of life and death, of creation and destruction. There is a like association of sculpture with temple and tomb. In tropical countries of teeming life, vegetable, animal and human, it would not be surprising to find a corresponding richness of images. They are profuse in the art of India, though India also provides the essentially simple image of calm and philosophic wisdom in the figure of Buddha. The Hindu religion personified the forces of nature in many different forms, as for instance in the "tree spirits" such as are to be found in the gateway sculptures of the shrine of Sanchi, South Bihar, executed in the first century B.C. Buddhism, which became defined as a system of enlightenment and detachment from worldly concerns in the third century B.C., concentrated visual attention on the ideal human figure.

As the Greeks had conceived in sculpture the idea of the physically perfect being, so the Buddhists conceived the idea of a spiritual perfection. The two conceptions approach one another in the calm of feature and serenity of pose. It is one of the most dramatic aspects of the history of sculpture that the connection was one of actual influence as well as of parallel lines of thought. Alexander the Great, in his invasion and attempted colonization of northern India, 327–5 B.C., brought with him Greek artists and examples of Greek art. Their influence is to be found in the Gandhara school of the North-West. It remained active in the early years of our Christian era. In the reign of the Guptas, A.D. 320–470, the Indian sculptors had evolved the typical Buddha figure in whose features there remained a distant likeness to Apollo. For a later Buddha figure see Plate 19.

The spread of Buddhism throughout the Far East is marked by a corresponding development of sculpture. Varying somewhat in conception according to the nature of the countries, the Buddha figure remains recognizably distinct in the main regions of what is now Indonesia (Sumatra, Java), Viet Nam (Tonking, Annam, Cochin China), Laos, Cambodia, Thailand, China and Japan. The ancient capital of the Khmer Empire (Cambodia), Angkor, has among its astonishing monuments huge sculptured heads (end of the twelfth century) representing an incarnation of the Buddha in the king. The Thais in Siam produced many beautiful examples in a clear-cut style of their own, the fourteenth to the fifteenth century being a flourishing period. The idea and the sculptural conception travelled eastwards along the main route of communication into Chinese

Turkestan and China itself. The Buddha is carved in the grottoes of Yun-Kang, northern China (fourth to sixth century A.D.) and monumentally farther south in the grotto of Long-Men (seventh century). This period saw also the appearance of the Buddha in Japan.

There were, however, many other developments in sculpture apart from the remarkable progress of this religious idea. In India the Hindu gods, Vishnu, Brahma and Shiva, creator and destroyer, were the subject of such works as the colossal figures carved from the rock in the island of Elephanta, off Bombay. Shiva, many-armed and dancing, is the subject of magnificent bronzes produced in southern India. Figures carved in sensuous rhythm, dancing and making love, are remarkable Hindu works in the caves of Ellora, Hyderabad, dating from the eighth century.

In Java, in the period when Barabadour was the capital (from the eighth century), sculptors carved elaborate reliefs. There is a teeming profusion of descriptive detail in the bas-reliefs of Angkor Vat (twelfth century) recording the triumph of the ruler Souryavarman II, on elephant back with his troops around him.

The sculptural genius of China is perhaps more characteristic in works of a different kind from the Buddhist sculptures. Some show an imaginative feeling for the grotesque. It appears in representations of the lion, an animal unknown in China but having a distant origin in the Asiatic lion so often a subject of Mesopotamian and Iranian art. It was converted by the Chinese sculptors into a truly fabulous creature. A wonderful sense of animal life, however, directly observed, appears in the reliefs and ceramic sculpture of the T'ang and Sung dynasties (7th–13th centuries A.D.) The horse is modelled with an observation and grasp of essentials of form that recalls the horses of the Parthenon.

A feeling for the beauty of material and an abstract refinement of form are found in the small objects of jade or soapstone, emblems of good fortune or else purely ornamental. There is some likeness in small sculpture between China and Japan. The idea of combining art and use, a strong element in Japanese art, created a special link between sculpture and what is known as "applied art". It is illustrated by the ingenious variety of carving devoted to the *netsuke*, the button fastening, in wood or ivory, of the girdle, to which small objects could be conveniently attached.

CHAPTER VII

THE GROWTH OF SCULPTURE IN THE CHRISTIAN WORLD

THE break-up of the Roman Empire inevitably produced, with all its great and various changes, a new direction in sculpture. The two facts that stand out are, first, that sculpture gained a fresh life with the rise of Christianity and, second, that a separation of the Christian East from the Christian West caused it to branch out in somewhat different ways. The new Rome established on the shores of the Bosphorus in the fourth century, the old Byzantium turned into the city of Constantine, Constantinople, was distinct from the atheist empire of the Caesars in having adopted Christianity as the official religion. This, sculpture, like the other arts was henceforward mainly to serve.

Byzantine sculpture is a complex subject, however. The new empire was conservative, desiring to maintain classical tradition in art as in other respects, though departing from it even unconsciously. For a long time the departure was regarded as "decadence", one aspect of the general "decay" which the author of *The Decline and Fall of the Roman Empire*, Gibbon, regarded as "premature and perpetual". That it was a mistaken view is now plain. Constantinople has been well described as "the Paris of the Middle Ages", a rich and civilized city where an exquisite art flourished.

As far as we can tell, though the subject does not seem to have been exhaustively studied, there was little free-standing sculpture on a large scale, such as had been devoted to the pagan deities. Some portrait sculpture of the classical kind remains from the early Byzantine centuries. But the principal works are exquisite reliefs on ivory book covers, caskets and plaques, with representations of Christ, the Virgin and Saints, and also small objects and figures, fashioned in the precious metals. The art of carving attained its peak between the sixth and twelfth centuries A.D. A masterpiece is the ivory "Harbaville Triptych" (Louvre) dated to the second half of the tenth century (Plate 32). Many other superb examples are in the Victoria and Albert Museum. A masterpiece in another form is the silver gilt relief of Virgin and Child (eleventh to twelfth century) in the Cathedral of the Venetian island of Torcello.

Various influences of style appear. Such Byzantine dependencies as Alexandria and Antioch (until their conquest by the Moslems) were centres of ivory carving, and oriental characteristics have been discerned in examples whose origin is uncertain. Yet the Hellenistic feeling of the Greek artists who found employment at Constantinople and elsewhere seems dominant in the idealism and refinement of the figure sculpture.

It is not possible to make an absolute division of Byzantine from sculpture of similar date in the more Western lands, for the influences of art are carried abroad and germinate like seeds. Yet broadly speaking it can be said that early Christian art in the West was more inclined to follow the narrative style of relief such as the Romans had used on their triumphal columns. An example is the "Basilewsky Situla" (Victoria and Albert Museum), an ivory bucket for holy water, tenth century, carved with scenes of the Passion and the Resurrection, the figures being crowded and compressed into the compartments of the surface. There is a Christian sarcophagus of the fourth century in the Lateran Museum at Rome which represents, in this style, the story of Jonah and the whale.

In the north of Europe, in the "Dark Ages" of migration, conflict and a gradual return of order, there was a fusion of pre-Christian forms of sculptured ornament. A Northumbrian chest of *c.* 700 (British Museum) shows scenes of pagan mythology and Christian imagery together, carved in whalebone. A dragon and snake on a stone of the eleventh century, preserved in the London Guildhall collection, gives an example of the interlaced ornamental conventions of pagan Scandinavia. The style is Christianized in the interlaced ornament of the crosses for which Northumbria was notable in the eighth century.

Several factors made for a more intensive cultivation of sculpture in the West than in the Byzantine world. The fear of idolatry caused the Byzantine Emperor, Leo Isauricus, to ban the use of images by the Church early in the eighth century and though the prohibition was later rescinded an attitude remained, averse from free and naturalistic representation in sculpture. In the West, however, sculpture was an expanding art in association with the programme of church building which brought a civilized order again into what had been the old Roman provinces. The result was the style called Romanesque which reached its height in the eleventh and twelfth centuries, and is found in France, the Rhineland and Britain. On a small scale one can see the influence of Byzantine sculpture, as in an Anglo-Saxon masterpiece of the eleventh century, the "Adoration of the Magi" carved on whalebone (Victoria and Albert Museum). (Plate 33). The massive architecture of the period however made for a broader and more monumental form of carving to go with it. The Roman style of representing incidents also appears in the reliefs which served an instructive and awe-inspiring purpose.

Cathedrals and other churches, Chichester in England, Hildesheim in Germany, and a great many in Burgundy and northern Spain, contain remarkable works of Romanesque sculpture. A subject often found is "The Last Judgement", as an elaborate and dramatic reminder of awards and penalties, carved in the semicircular arch over the cathedral porch. A superb late example is that of the Burgundian cathedral of Autun, 1130–40, of which the sculptor is known by name, Gislebert. It combines the majestic figure of supreme authority with a vivid rendering of the happiness of the elect and the fate of the damned.

In the early Middle Ages there had been a haunting idea that the Last Judgement would come with the millennium, i.e. A.D. 1000. A plausible connection has been made between the feelings of relief and hope which grew after the fateful date had gone by, and a great change in art. Though the Last Judgement remained a subject for sculptors it was now simply a general and undated warning. Otherwise in sculpture there was a new feeling for humanity and nature.

What is known as "Gothic" sculpture, like the Romanesque, was closely connected with architecture. It developed with a great change in architectural construction. The brilliant engineering which produced a new system of distributing weight, dispensing with the ponderous mass of Romanesque masonry and enabling the lightened fabric to soar gracefully upwards, had two effects on sculpture. It provided a new variety of surfaces and recesses in which it might appear. It made also for a change from the horizontal to the vertical which released figures from the cramping and rather stunted character which had prevailed since Roman times. The north of France in the twelfth and thirteenth centuries was the great fount of style. The cathedrals of Chartres, Reims and Amiens had their wonderful array of saintly and angelic figures, the elongated forms seeming to convey ideal aspirations, the features carved with a warmth of sympathy for beauty of expression. One of the sculptors' workshops at Reims produced that famous figure of a smiling angel on the east porch, 1260–80, which is one of the supreme Gothic masterpieces (Plate 35).

As the Church was international so was the style of architecture and sculpture (contemptuously described as "Gothic" by a much later age). The extraordinary profusion of sculpture which accompanied the great period of cathedral building is to be found in every part of western Europe. It extends in date from the end of the twelfth to the fifteenth century. Broadly speaking, thirteenth-century work is marked by the spiritual dignity of figure sculpture. Beautiful examples in England are the reliefs of angels in the south transept of Westminster Abbey, *c.* 1250–5, the statuary of the west front of Wells Cathedral, and the figures of the Judgement Porch, *c.* 1260, of Lincoln Cathedral. An ivory diptych of the Virgin and Child and Christ (Victoria and Albert Museum) retains the nobility of architectural sculpture on a small scale.

The fourteenth century was less austere and showed a growing delight in nature, as for instance in the treatment of the capital of a column which, instead of a formal pattern, as in the classical or Byzantine capital, became a naturalistic carving of foliage. What remains of the tomb of St Frideswide in the Cathedral, Christ Church Oxford (end of the thirteenth century) shows such carvings of bryony, hawthorn and oak leaves. Imagination and humour had freer rein. Demons and monsters were imprisoned as it were in stone. The new system of draining the roof that was now introduced, by a gutter pierced with projecting spouts of stone, created that special product of Gothic fancy, the gargoyle. The brackets on the underside of the choir seats, which when turned up provided a measure of support for the standing worshipper, became a vehicle for the whimsical carver of misericords. Biblical incidents, satirical subjects, incidents of fable and legend, and grotesque animals were treated in an extraordinarily inventive and often humorous way. English churches, Beverley, Ely, Gloucester, Lincoln, Ripon, Worcester—there is a long list—are rich in this form of expression by men who would be known as carpenters rather than sculptors but were certainly natural artists.

As always in the history of sculpture, the tomb and the funerary portrait had their special forms. The royal portrait, as in earlier ages, was made serene and godlike. The statues of Henry III and Eleanor of Castille in Westminster Abbey, cast in bronze by the London master-goldsmith, William Torel, in 1292, give an English example (Plate 36). The knights were portrayed in the full armour of chivalry—every detail of the Black Prince's gorget of chain mail is picked out in the impressive effigy in Canterbury Cathedral (1376). An English masterpiece of the fourteenth century is the canopied Percy Tomb in Beverley Minster, *c.* 1345, commemorating Eleanor, wife of Henry Percy. The work of the school of sculptors (or "imagers" as they were called) centred at York, it combines grandeur with a lavish wealth of ornamental detail.

In northern Europe, sculptors continued to work in the Gothic spirit during the fifteenth century. A great number of religious images in three dimensions were carved in stone or wood in Germany and the Netherlands. Pride in craftsmanship was often more in evidence than the ideal feeling of earlier Gothic. It appears in the elaborate treatment of drapery, so often imitated from sculpture by the Netherlandish painters of the century. Distinctive enough in style to be known by name are such late Gothic masters as Claus Sluter, Hans Multscher and Tilman Riemenschneider, whose masterpiece is the carved wood altar at Creglingen. There was Bernt Notke, noted for the ornate and realistic St George at Stockholm, and Veit Stoss whose "Angelus" in the church of St Lorenz, Nuremberg, was described by Europe's first art historian, Vasari, as "a marvel in wood". In England Henry VII's chapel in Westminster Abbey brings into combination English sculpture, still carried out in the Gothic style, and the work of an Italian artist, Torrigiani, trained in the very different tradition of the Renaissance (Plate 37).

CHAPTER VIII

RENAISSANCE, BAROQUE, ROCOCO

THE course of sculpture in Italy was separate, even in the Middle Ages, from that in northern Europe. There was no drastic change in architecture such as produced the Gothic cathedral and sculpture in close relation with it. The links with the classical past were closer. From one point of view the Renaissance, as the word implies, was a rebirth, a return to the ideas and achievements of ancient Greece and Rome in art. In that sense it had no religious base. At the same time it was a new idea of the capacity of the human being to think and act in a way that made for the full development of his powers.

The link with the past appears in the work of Niccolò Pisano. He lived in the thirteenth century at a time when the northern cathedral sculptors were producing their Gothic masterpieces, though he had nothing in common with them. He studied and tried to rival the ancient sarcophagus reliefs he saw at Pisa. Christian subjects and antique manner come together in his masterpiece, the pulpit of the Baptistery of Pisa, *c.* 1260. His son Giovanni and pupil Andrea who took the name of Pisano carried on his work. Some works by Giovanni Pisano show that emotional force in which Italian sculpture was to go beyond the limits of classical art (Plate 34).

The typical Renaissance sculptor appears in Lorenzo Ghiberti, whose bronze gates to the Baptistery at Florence are one of the great triumphs of Italian art. Ghiberti is in many ways the first typical sculptor of the Renaissance. He stands out as an individual known by name and personal achievement, a fact which distinguishes sculpture, from the fifteenth century onwards, from the work of the anonymous craftsman in the humble service of Church or State. He studied the ancient Roman art of narrative relief. Like other Italian artists of his time, he was also filled with the desire for scientific knowledge making for a greater command of realistic effect, perspective in particular occupying his attention. The panels for his bronze gates were in two series from the Old and New Testaments. They were cast separately and then fitted into place in the door. Architecture, landscape and figures in the various scenes were composed with wonderful skill, differences in the depth of relief giving a remarkable illusion of space. Considered by Michelangelo worthy to be the gates of Paradise, they took the greater part of the artist's lifetime to complete (Plate 46).

Another great sculptor who devoted many years of his life to a series of church reliefs was Jacopo della Quercia. His work at San Petronio de Bologna was characteristic of the new race of sculptors in its individuality. In complete contrast to the elegance and richness of detail in Ghiberti's work and its pictorial interest, Jacopo della Quercia conceived a

"Temptation" or a "Presentation in the Temple" in boldly simplified forms. In concentrating on the expressiveness of the human figure and avoiding other detail he already foreshadows the work of Michelangelo.

The expression of power, energy and emotion through the human figure, anatomically mastered as it had never been in the past, was the special achievement of the great Renaissance sculptors. They modelled nude figures with a freedom never allowed to the Gothic artist, the more freely because their cultured secular patrons delighted in subjects taken from classical mythology. Though many great works continued to be carried out in relief, the revival of free-standing sculpture in the open was notable.

This was one of the achievements of Donatello. His equestrian statue of the military commander, Gattemalata, at Padua, is a masterpiece in a type of sculpture in which the Italians excelled. In grandeur of pose it has a rival only in the equestrian monument to Bartolommeo Colleoni at Venice by another Florentine, Andrea Verrocchio (Plate 39). Donatello was an artist of varied moods, sometimes gentle, sometimes dramatic in feeling and harsh in style (Plate 42). Among his statues in the round that of the youthful David is full of grace and charm, his "Judith and Holofernes" sombre. He was equally a master in marble and bronze and, through his pupils and example, created the tradition of the bronze statuette which was brilliantly carried on in the fifteenth and sixteenth centuries in Florence, Siena and Padua.

Donatello had famous successors at Florence in Pollaiuolo and Verrocchio. They reveal another characteristic of the Renaissance in being all-round artists equally ready to undertake commissions for sculpture, painting or goldsmith's work. Antonio Pollaiuolo, who excelled equally in painting and sculpture, is notable in having used the study of anatomy to the purpose of representing such vigorous movement as appears in his "Hercules and Antaeus".

Many beautiful works in bronze were produced by other Renaissance sculptors—Francesco di Giorgio, Bertoldo, Rustici, Bellano, Riccio (Plate 49), Antico, all of whom were considerable masters. Classical themes were the subject of most works on a small scale. Another development was that of ceramic sculpture as practised with so much charm and attraction of colour by the Della Robbia family (Plate 47).

The expression of energy and power as an aim of art is to be found in the two greatest artists of comprehensive gifts in the Renaissance age, Leonardo da Vinci and Michelangelo. No work of sculpture authentically from the hand of Leonardo da Vinci exists. The equestrian monument he designed to Lodovico Sforza at Milan got no farther than the model which was destroyed. Yet there are statuettes of rearing horses, possibly made after the wax models he produced for the fresco of the Battle of Anghiari (which never came to fruition). The celebrated "Riding Warrior" in the Budapest Museum conveys the magic of genius which all his work possessed in its incomparable force of movement (Plate 38).

This dynamic force attains its full expression in the work of the greatest of sculptors, Michclangelo. It was the product not only of an immense technical skill, but of a great mind, seeking, alike through classical and Christian themes, to convey the capacities inherent in humanity. The restless thought and discontents of the artist himself seem to generate a vital current in the marble he carved.

Michelangelo as a boy, under the patronage of the Medici, was trained by Donatello's pupil, Bertoldo. His earliest known sculpture, before he was twenty, is a Madonna in shallow relief after Donatello's manner. A relief executed soon after was antique in theme. "The Battle of the Centaurs", which shows figures in violent action rather after the style of a battle relief by Bertoldo (preserved in the Museo Nazionale, Florence). Early masterpieces reveal a conflict of ideas. The nude statue of a soft and self-indulgent Bacchus (Museo Nazionale, Florence) is an adaptation of the antique, executed when he was twenty-one. Very different in feeling is the austerity and sadness of one of his most beautiful works, carried out shortly after, the "Pietà" in the Basilica of St Peter. He returns to the antique in the life-size "David" carved from a single block of marble (Florence, Accademia), the ideal figure of a young athlete in which the sculptor proudly displays his mastery of anatomy. Yet soon after he is engaged on reliefs of the Virgin and Child of which the "Taddei Madonna" (Royal Academy) is an example (Plate 44).

His giant abilities were desperately engaged against many difficulties in the effort to produce the grandest of tombs for Pope Julius II. There remain only magnificent fragments of the original scheme—the series of supporting figures, "slaves" or "captives", some of them none the less impressive for being only roughed out. In the final simplified and imperfect version of the tomb (Rome, S. Pietro in Vincoli), there is the majestic figure of Moses to suggest what it might have been. The greatness of his conception was more completely realized in the tombs of Lorenzo and Giuliano de Medici (Plate 43). The reclining figures of Day and Night, Dusk and Dawn are among the most superb creations of all sculpture.

It was remarked by his contemporary, Vasari, that Michelangelo left more statues unfinished than finished. In some works this was deliberate. He intended to give the dramatic sensation of a figure struggling to emerge from the marble that contained it. Sometimes also he made a special point of contrasts of surface, as between the parts that still bore the rough marks of the chisel and those which were polished so as to catch the light. The method is used with extraordinarily moving effect in Michelangelo's last work, the "Rondanini Pietà" (Milan, Castello Sforzesco) (Plate 45). For sculpture which aimed so intensely at conveying ideal aspirations in humanity it was perhaps necessary to be incomplete.

It would evidently be a mistake to think that the Renaissance sculptors aimed at the same kind of perfect beauty and physical equilibrium as the Greeks of the classical period. The image of human power, the violence of expression that is found in the works of such giants as Donatello and Michelangelo was the result of a different quest, a "divine discontent". The restlessness, the tendency to exaggerate and distort, becomes pronounced in the later Renaissance period. They are the principal features of what is known as the Mannerism of the sixteenth century.

Mannerism was largely the result of an admiration for Michelangelo which came near to idolatry. It was sometimes also the result of envy like that of his imitator, Baccio Bandinelli. The dominant figure of the sixteenth century was a northerner who worked in Italy, Giovanni da Bologna, more than an imitator, a man of imagination. His work strongly reflects the taste of the time for the classical themes which allowed of a free treatment of the nude figure in action, such as Mercury in flight or the "Rape of the Sabines" (Plate 52). He brings into view also the value of sculpture in a secular and decorative aspect, as in the structure of a fountain. One of his principal achievements is his fountain at Bologna, at once architecture and sculpture, surmounted by a huge figure of Neptune. His imagination appears in a gigantic figure of Jupiter, in the Medici gardens at Pratolino, crouched on the back of a marine monster from whose mouth a stream of water flows into a pool.

Benvenuto Cellini reminds us of the close relation of Renaissance sculpture with the craft of the goldsmith and metal worker. A great amount of beautiful work on a small scale was produced in Florence, Venice and Padua in the sixteenth century by many able artists, Niccolò Tribolo, Vincenzo

Danti, Bartolommeo Ammanati, Jacopo Sansovino, Danese Cattaneo, Alessandro Vittoria and others. Sometimes they were small bronze copies of marble statues, or studies cast in bronze for larger statuary, or, again, ornamentally figured articles such as a salt-cellar, an inkstand, a door knocker, a candlestick, a furniture mount. Cellini is the most famous of the sculptor-goldsmiths. His statue of Perseus with the head of Medusa, 1533 (Loggia dei Lanzi, Florence) can be numbered among the great works of the Italian Renaissance (Plate 48). The elaborate salt-cellar of solid gold he made for Francis I of France was a piece of sculpture as well as of craftsmanship. He was one of the Italian artists who introduced a mannered style of Renaissance sculpture into France.

The splendour of Renaissance sculpture in Italy, causing its influence to spread abroad, marks the end of the Gothic style where it still persisted. This was so in France where Francis I invited Italian sculptors and painters to his court. An elegant style of stucco sculpture was used by Primaticcio as a decorative adjunct to his paintings in the palace of Fontainebleau. His figures show the Mannerist exaggeration of limb with graceful effect. The "Nymph of Fontainebleau" (1542), a high relief in bronze, made for a doorway of the palace by Cellini, shows a similar elongation of figure. Not certainly attributed to any one artist but in the Franco-Italian manner is the group of Diana with a Stag, 1558–9 (Louvre). A great French sculptor who gives to this direction of the art a fresh and original beauty is Jean Goujon. Rhythmic

33. BYZANTINE (*left*): *Leaf of Ivory Diptych* (*Symmachi*). *Fifth century.*
(*Victoria and Albert Museum*)

ANGLO-SAXON (*right*): *Adoration of the Magi. Ivory. Eleventh century.*
(*Victoria and Albert Museum*)

34. NICCOLO and GIOVANNI PISANO: *Pulpit, Siena Cathedral, 1268*

35. GOTHIC: *Detail, the "Bel Ange". Thirteenth century, Reims Cathedral*

36. WILLIAM TOREL: *Eleanor of Castille, Westminster Abbey, 1291*

37. PIETRO TORRIGIANI: *Effigy of Henry VII, Westminster Abbey, 1512–18*

38. LEONARDO DA VINCI (after): *Riding Warrior. Bronze.* c. *1506–8*
(*Budapest Museum of Fine Arts*)

39. ANDREA VERROCCHIO: *Equestrian Statue of Bartolommeo Colleoni, 1479–88, Venice*

40. AFRICAN: *Ife "Classical" Head*
(*British Museum*)

41. AFRICAN:
Two Ivory Leopards with Copper Decoration, Benin

(*Reproduced by gracious permission of Her Majesty The Queen*)

(*British Museum*)

42. DONATELLO: *St George, 1412–15*
(*Museo Nazionale, Florence*)

43. MICHELANGELO: *Tomb of Giuliano de Medici*, 1524–33, *Florence, Basilica of San Lorenzo*

44. MICHELANGELO: *Taddei Madonna. Unfinished Marble*, c. 1505–6
(*Royal Academy*)

45. MICHELANGELO: *Rondanini Pietà. His last work*
(*Milan, Castello Sforzesco*)

46. LORENZO GHIBERTI: *The Creation of Adam and Eve. Panel of Doorway. Early fifteenth century, Florence, Baptistry*

47. LUCA DELLA ROBBIA: *Boys Singing and Playing. Detail of Cantoria, 1431–40*
(*Museo di S. Maria del Fiore, Florence*)

48. BENVENUTO CELLINI: *Perseus with the Head of Medusa*, c. 1540
(*Loggia dei Lanzi, Florence*)

flow of drapery and grace of pose in his reliefs for the Fontaine des Innocents, Paris (Louvre) make it the first of many works entirely French in feeling (Plate 53). After more than three hundred years it inspired in the young painter, Auguste Renoir, a sense of the way in which feminine beauty could be translated into art by an artist of a later time.

In Germany there was a more gradual intermingling of medieval craft and Renaissance style in the production of elaborate carved altars and reredoses. In England the Renaissance influence is principally marked by the bronze figures for the tomb of Henry VII, executed by the Florentine, Torrigiani (Plate 37). The Renaissance style was propagated in Spain by the sculptor, architect and painter, Alonzo Berruguete. Yet the sculpture of the Italian Renaissance was on the whole too strongly individualistic to become an international style as Gothic had been. It was the later development known as Baroque which attained this result.

As in other ages the motive behind this new impulse, which gained momentum at the end of the sixteenth and during the seventeenth century, was a reassertion of authority in Church and State. Architecture, sculpture and painting alike were again regarded as the propaganda of a faith and of a ruling power. Sculpture was one of the instruments of the Catholic Church against the separatist movements of the Reformation. In its Baroque form it was an art of the Counter-Reformation. It flourished in Italy, Spain, France and in the Catholic princedoms of the Teutonic lands. It extended to the Spanish possessions in the New

World. It found less access to the Protestant and Puritan regions of Europe.

In several ways the Baroque style was an extension of that of the Renaissance in its full development. The sculptor had at his command the mastery of movement and gesture bequeathed by Michelangelo in both plastic and pictorial masterpieces. It was used now with a deliberate effort to play on the emotions of the spectator, to create a kind of rapturous astonishment like that of some elaborate operatic performance. The likeness was pronounced, inasmuch as architecture, sculpture and painting were united in the same way as opera combined music and spectacle, colour and song. The whole impression was one of theatre, with daringly contrived effects of fluttering figures, emphatic attitudes, illusions of movement in space and accumulations of detail with an overwhelming effect on the senses.

As in the Gothic period (and in contrast with the Renaissance period) the individual artist was again subordinated to an overriding purpose. A good deal of the architectural sculpture of the Baroque age is anonymous. Yet the spirit of the enterprise can be illustrated by the work of an artist of astonishing abilities, Giovanni Lorenzo Bernini. In one phase he may be compared with Giovanni Bologna, in the creation of giant fountain figures among tumbled rocks like those of the Piazza Navona and the Fontana di Trevi at Rome, or in classical legend conceived in terms of animated movement like his "Apollo and Daphne" and "Pluto and Proserpina" (Rome, Borghese Gallery). He created a style of portraiture as grandiose as the bust of Louis XIV at Versailles, as elegant as that of Mr Baker (Victoria and Albert Museum), whose collar is a marvellous counterfeit of lace in carving. Above all it is in his work for the Church that he is most amazing.

His sense of the dramatic appears in the elliptical colonnade of the piazza before St Peter's, with its 162 Baroque statues of saints; in the huge bronze canopy or Baldacchino of the interior, 95 feet high, with its gilded spiral columns; the ornate tombs of Paul III and Urban VIII; the bronze throne of the Tribune surmounted by angels hovering among clouds. The melodramatic appeal to the emotions reaches its height in the swooning ecstasy of St Theresa in the carved altarpiece of S. Maria della Vittoria, Rome (Plate 50).

Other brilliant sculptors of the Baroque period in Rome were Alessandro Algardi, Melchiore Caffa and Giuseppe Mazzuoli. Bernini's pupils were numerous. His influence was strong in the France of Louis XIV, the Baroque style being well adapted to convey the pomp and authority of the court of the Grand Monarque. It is represented at Versailles by the work of Pierre Puget. His sculpture presents a contrast with that of François Girardon, also employed at Versailles; for Girardon still adhered to the ideal of feminine charm which is represented by the earlier school of Fontainebleau.

Palaces and churches in Germany and Austria became vehicles of the Baroque style in the early eighteenth century. In Germany, with its multitude of local rulers (as many as 350 minor dynasties), there was a craze for building Baroque palaces, adorned with a fantastic amount of sculpture, over-lavish and fussy in effect. There was a corresponding tendency in church architecture as in the "Asam" Church at Munich where the brothers Kosmas Damian and Egidius Asam, contrived the most startling effects of angelic figures and cherubim poised in space.

The disapproval of Protestant and democratic lands for Baroque sculpture as being falsely artificial and deplorably melodramatic was only to be expected. It made clear the opposition in religious and political views. In England, Puritan zeal in the seventeenth century had led to the destruction of medieval statues in churches as idolatrous. Suspicion of a style so much associated with an alien Europe as the Baroque was the more intense. It was one of the indictments against Archbishop Laud that he had allowed to be erected at Oxford the Baroque porch of the church of St Mary the Virgin. This beautiful work by Nicholas Stone, with its Berniniesque twisted columns and statue of the Virgin and Child, was considered an offence. Yet style in art had its way of developing independently of prejudice on religious grounds. Sir Christopher Wren had an eye for the dramatic value of Baroque in designing St Paul's Cathedral. Its expression in architecture is emphasized by the statues on the exterior by Francis Bird and the ornamental detail of Grinling Gibbons.

A persistent form of Baroque sculpture in the eighteenth century was that of the church monument or memorial. It was a development of the Renaissance tomb in which an architectural framework and allegorical figures made an impressive setting for the portrait image. The Medici tombs of Michelangelo give a great example. The spread of the style in Europe may be illustrated by the monument to the Duke of Lennox and Richmond in Westminster Abbey, designed by the French sculptor Hubert Le Sueur. Le Sueur (whose work includes the equestrian statue of Charles I in Trafalgar Square) places over the sarcophagus an elaborate structure supported by figures representing Faith, Hope, Charity and Prudence, the whole being surmounted by a figure of Fame.

The Baroque tomb added the theatrical exaggeration typical of the style in general. The allegorical figures took on movement. Symbolic objects were added in profusion. Symmetry was abandoned in favour of an irregular composition used to heighten the emotional effect. Uneven shelves of rocks, flowing draperies, emblems and trophies, architectural details were combined with prodigious skill. Puritanical disfavour was aroused no less by this form of Baroque than by others. It is only in comparatively recent times that it has been impartially viewed as an art form.

The monuments in Westminster Abbey provide some remarkable English examples. They are in contrast with the Gothic architecture that is their background, to such an extent that the eye cannot take in both together without a discomfort produced by their discrepancy. Yet considered in themselves and making as much allowance for personified qualities and "conceits" as we do in the poetry of the period, they sometimes attain a daring brilliance. François Roubiliac stands out as the master of the Baroque tomb in England. The monument to the Duke of Argyll (1748–9), with its figures of Liberty, Eloquence (rhetorical in gesture), Minerva and Fame (inscribing the nobleman's titles on her scroll), is one of his principal achievements. Another famous work by him in the Abbey is the Nightingale Monument (1761) (Plate 51). It records the fate of a Mrs Nightingale who died of shock after a lightning flash had struck. Death, as a grim skeleton coming up from below, points his lance at the lady whom her husband vainly tries to shield.

On the whole, however, the eighteenth century was losing both the active spirit of religious propaganda, which had made sculpture an instrument of the Counter-Reformation, and its character as an expression of sovereign power such as it had in France during the reign of Louis XIV. The Rococo style which followed the Baroque and was centred in France was lighter, though no less elaborate, more charming than theatrically impressive, more decorative, in the sense of being allied to the secular furnishing of an interior. It was a playful rather than an heroic or monumental style and in that spirit made use of the classical repertoire of nymphs, satyrs, and cupids with elegant effect. Love in various guises was a favourite theme as in the work of Nicolas François Gillet, of Claude-Michel Clodion, of Etienne-Maurice Falconet. Groups such as those by Falconet of "Venus chastising Cupid" and "Cupid and Psyche" (there are examples in the Wallace Collection) were popular. Bronze sculpture became a form of "applied art", adding to the rich effect of pieces of furniture. The bronze figure as an adjunct of a clock or a lampstand became a tradition which lingered on as a debased and redundant style of decoration into the nineteenth century.

Rococo sculpture reflected the character of a frivolous court. A greater development was that of portraiture as represented by Jean-Antoine Houdon. France had previously excelled in this branch of sculpture. The court sculptor of Louis XIV, Antoine Coysevox, produced impressive works in a grandiose manner influenced by Bernini. It was Houdon's achievement to bring out the humanity rather than to stress the formal grandeur of his subjects. His activity was not confined to the French court. He travelled widely, being invited to America to make the statue of George Washington (a replica of which stands in front of the National Gallery, London). Masterpieces by Houdon are the marble head of Voltaire, 1781, placed in the foyer of the *Comédie Française* (Plate 52), the satirical smile of the philosopher being vividly rendered, and the charming bust of Mme de Serilly, lady-in-waiting to Marie Antoinette (Wallace Collection).

CHAPTER IX

THE NINETEENTH CENTURY

TOWARDS the end of the eighteenth century there was a reaction against the frivolity of the Rococo style in art which took the form of a Classical Revival. In France, which had been the great centre of Rococo art, it was an aspect of the revolutionary feeling that was to do away with the monarchy and aristocracy. The luxurious art which had flourished under their patronage was now the object of dislike. Yet the movement was far from being limited to France, and did not necessarily have the political implication that it had in the paintings of the French artist Jacques Louis David. It was an attempt to "purify" sculpture by a return to the noble simplicity of ancient Greece. It grew up in Rome, the seat of classical tradition, the main repository of the copies of Greek masterpieces. A leader of the movement was the Italian sculptor, Antonio Canova. His colossal marble group, "Theseus and the Centaur", 1782 caused him to be hailed as the founder of a school. A nude statue of Napoleon, posed like a classical deity or hero and holding the emblem of victory (Brera, Milan), represents his leaning towards the antique.

The attempt to rid sculpture of the ornate and voluptuous character it had acquired in the eighteenth century was congenial to the puritan spirit of northern Europe. Associated with the classical revival are the British sculptors, John Flaxman, his pupil, Edward Hodges Baily, and John Gibson; the Danish sculptor of Icelandic origin, Bertel Thorwaldsen and a number of other Scandinavians, Danish and Swedish. The French sculptor, François Rude, used classical features of style, though adding a rhetorical zest which appears in his "Marseillaise" group for the Arc de Triomphe, Paris (1836).

The sculpture of the classical revival tended to be cold and stilted in effect. This seems to have been due to its imitative nature. Perhaps the feeling of inferiority produced by the idea of vying with Phidias was, consciously or unconsciously, transferred to stone. In a way the Renaissance also had been a classical revival, yet it had never been insipid as the early nineteenth century often was. The reason, one might simply say, was that the Renaissance sculptors were greater men. Yet they had also a forward impetus, a new kind of mastery to achieve. The impetus was lacking in the nineteenth century. A great deal of sculpture on more or less classical lines was produced throughout the period but was, generally speaking, lacking in originality and animation. Contributing to this was the practice of employing workmen to reproduce a sculptor's clay model in marble, with a mechanical result.

Sculpture was no longer a corporate effort as it was in the Gothic or even in the Baroque period. It had nothing to convey as in the past on behalf of Church and State. It was an individual expression. Varying in quality, it was at its best when freely and individually expressive. One of the most interesting features of nineteenth-century sculpture is the work of those who were primarily painters and freer and more experimental than the average professional sculptor. Géricault, romantic genius of painting, was one of the first to turn to sculpture and to give it a new energy of form. As the old masters had done, the caricaturist Daumier made models which provided suggestions for the light and shade of his lithographed caricatures. These are remarkable for their unconventional treatment and the powerful sensation of character that they give.

A painter also, Antoine Barye was one of the greatest of animal sculptors, though his aims were very different from those of the ancient artists for whom the animal was an enemy to be destroyed. There was now a reflection of the romantic delight in the freedom of the wild and in the spectacle of its untrammelled energy. Like Delacroix in painting, Barye in sculpture conveys the primitive force of nature in his representation of nature red in tooth and claw, the lion fighting a serpent, the jaguar devouring a hare (Plate 49).

Two great French painters in the second half of the nineteenth century turned to sculpture with striking result. Edgar Degas, when failing sight made painting difficult, took to modelling (he called sculpture "a blind man's art", meaning that it was essentially concerned with touch). His wax models were mostly made for his own satisfaction, only one being exhibited in his lifetime. Bronzes were made from those found in his studio after his death. They are mainly studies of movement, either of dancers or racehorses and full of vitality (Plate 53).

For the different reason that arthritis made it hard for him to hold a brush, Auguste Renoir in old age employed a young sculptor to carry out his conceptions in plastic form. They were evolved from sketches by him and the actual work was carried out under his close supervision. His bronze "Venus Victorious" might be called "classical" in the sense of being an ideal nude figure, but, like his late paintings, it is imbued with a living warmth which places it in a separate category from the rather frigid products of "classicism" as the term was usually understood.

In England also the painter-sculptors stood out. A remarkable isolated figure was Alfred Stevens who had something of the diversity of talents, as sculptor, designer and painter, which many artists in the Italian Renaissance had shown. In ideas and style he seems a throwback to that period, an artist born out of his time. His great work, on which he expended years of effort, was the Wellington Monument in St Paul's Cathedral. His feeling for design can be appreciated on a small scale in the bronze lions for the railings of the British Museum.

Painters who found a new stimulus in applying themselves to the problems of sculpture were G. F. Watts and Lord Leighton. Watts was a devoted student of the Parthenon sculptors but his sculpture, far from being imitative of ancient Greece, was surprisingly original. In his great work, the statue of "Physical Energy" for the Cecil Rhodes Memorial (in replica in Kensington Gardens), he achieved a dynamic combination of the movement of horse and rider, and the way in which the sharply cut planes carried out the effect was unusually bold for his time. "The Sluggard", by Lord Leighton, in which he caught the pose of a yawning model, was a work with an easy grace and balance in contrast with his more formal work as a painter.

It is France, however, that claims main attention in the nineteenth century, in particular by the towering figure of Auguste Rodin, who stands at the boundary of two periods, completing a tradition and also marking the beginning of a new age. Older contemporaries come near him in some respects, especially in the rendering of the supple movement and grace of women. J.B. Carpeaux was one of his precursors, a gifted artist who preserved the elegance of the eighteenth century in the time of the French Second Empire. Carpeaux's group "La Danse" for the Paris Opera House had a verve and flow of form which bring Rodin to mind. Another famous work, his "Ugolino" (1861) in spirit and attitude is close to Rodin's universally known figure of "The Thinker". Jules Dalou is another sculptor of an earlier generation than Rodin who provides a link between the elegance of the eighteenth century, as in his monument to Delacroix at Paris, and the work of his greater contemporary. Rodin, however, left both Carpeaux and Dalou far behind in the range of thought and power contained in his vast output.

His sculpture was no longer the work of the craftsman carrying out the behests of Church or State, but of the individual man of ideas and perceptions, capable of giving them powerful expression.

Rodin felt conscious of many links with the past, with Greek and Gothic sculpture, and (as another famous artist, Jacob Epstein, was later to do) he collected in the course of time a great many antique works. Yet he was perhaps most of all in tune with the individualism of the Italian Renaissance and in particular with the work of that other giant, Michelangelo. The deep impression made on him by Michelangelo's "Slaves" can be seen in a number of figures, which in the same way seem to struggle free from the block of marble. He makes much use of the same device of contrasting smoothly polished and roughly carved surfaces with dramatic effect.

Rodin's sculpture includes monumental works, figures symbolic of the progress of mankind, groups representing some poetic thought or emotional idea, portrait busts and statues, and a multitude of small figures, many of which were cast in bronze after his death. Like Ghiberti in the Baptistery gates at Florence, he planned a masterpiece in the doors he was commissioned to make for the Musée des Arts Decoratifs, on which he worked for more than twenty years. The subject, however, unlike the "Last Judgements" of the medieval cathedrals, was not orthodoxly Christian but represented his admiration for Dante. A tumult of images was inspired by the three parts of the poet's *Divine Comedy*, the *Inferno*, *Purgatory* and *Paradise*. So rich in incidental themes was the "Gate of Hell", which found its final home in the Rodin Museum, that Rodin treated a number of them separately and elaborated them on a large scale, an example being his famous group, "The Kiss".

Another great monumental work was the group of "The Burghers of Calais" (a version of which can be seen in the Victoria Tower Gardens, Westminster) in which he was inspired by the grandeur of Gothic sculpture, but into which he introduced a story element in a romantic fashion, more typical of his own time. Varied gestures convey the feelings of the six heroic citizens who offered their lives to Edward III if he would spare the town. One of his great symbolic figures was "The Age of Bronze", the figure of a youth representing the dawn of consciousness in mankind. His portraits show much psychological insight but his "Balzac", which goes beyond the exactness of portraiture, seems to infuse the stone with the power of some immense force of nature (Plate 55).

His small studies of female figures are remarkable for their free and unconventional rendering of movement, in which they may be compared with the figures of ballet-dancers in action which Degas modelled. Rodin's work was for a long time shocking to his contemporaries because of its daring originality. He was criticized in the reaction that followed the popularity of his later years on different counts. The stress on subject interest as in "The Thinker" and "The Kiss" seemed too obvious or too sugary for early twentieth-century taste, the execution of carved works, which he entrusted to assistants, giving some colour to the latter impression. The flowing and irregular movement of his figures was for a while considered formless. Yet Rodin's faults now appear trivial beside a greatness which opened up new prospects of expression.

CHAPTER X

SCULPTURE IN THE MODERN AGE

HAD everything it was possible to do in sculpture been achieved by the beginning of the twentieth century? It might have been thought that by then every possibility had been explored, that effort could be devoted only to imitating, as well as might be, the triumphs of earlier periods. There was, and still exists, a type of sculpture which tried to reproduce classical perfections of form, though for the most part in a rather tame and unconvincing way. Yet surprisingly there has come into being a development as original and powerful as any in the past.

This is in spite of the fact that the old close connection with architecture and, through architecture, with religious and secular authority, has declined. The connection has been closer with the independent and adventurous progress of modern painting and the ideas to which it has given rise. Sculpture has gained a new life in reflecting the thought, forces and problems of an industrial and scientific age like none that has gone before.

In studying its growth there is first to be considered the development at the end of the last century and the beginning of this which is known as *art nouveau*. Though mainly appearing in graphic art and objects of ornament, it appears in various arts as a style of mannered curvature. There is a trace of it in Rodin's flowing forms. In England the sculpture of Sir Alfred Gilbert represents it in brilliant fashion. A reaction quickly set in, however, against the elegant curvature in which there still lingered the Rococo style of the eighteenth century. It first showed itself among French artists and other artists in France. It was a desire for a stronger and more elemental basis of form which led them to look for examples outside the cultivated centuries of European art. The discovery that the productions of primitive races had qualities which made them admirable as works of art can be traced back to Paul Gauguin.

Gauguin's life among the natives of Polynesia and his interest in them not only led to many famous paintings but to a few sculptures carved in wood, which are remarkable not only in portraying native types but in a forceful simplicity of style, showing impressions derived from more than one source. These works date from about 1890. It was about fifteen years later that artists in Paris were excited by another discovery, that of native African sculpture. Its emphasis on basic forms and strongly marked features played an important part in the development of the most influential of modern movements in art, Cubism.

Reference has already been made to the way in which Modigliani used with great effect the same kind of simplified form that is to be seen in the African ritual masks. Otherwise the influence is first seen in paintings. It produced a new attitude towards the construction of objects in a picture, and the idea of an essential geometry in art which opened up an entirely fresh prospect. Cubism was a painters' movement. The painters led the way and the far-reaching effect of their work can be described by reference to painting alone. Yet an idea of art in which structure was so important could not fail to appeal to sculptors. The first of the Cubist sculptors was Alexander Archipenko, who produced remarkable works from about 1910. He made use of two methods that Cubism had introduced into painting. One was the emphasis on clear-cut geometrical shapes. Another was the idea that the same work might show different points of view of the object represented. In terms of three dimensions this entailed a new treatment of space. Archipenko made the hollows with which he perforated his statues into an active part of their effect. The aim was no longer that of giving a solid equivalent of the human figure but of attaining a new freedom and variety of form.

Perhaps this needs more explanation. It applied to sculpture an old axiom of painting that the "spaces left", the intervals between forms, are as important to the effect as the forms themselves. The result, translated into three dimensions, was to give a sense of inner structure and not only of the surface of a mass. A piece of sculpture became like a work of architecture in the sense that architecture may be defined as "the organization of space".

Other sculptors who gained inspiration from the Cubist painters were Henri Laurens, Jacques Lipchitz, Ossip Zadkine, Raymond Duchamp-Villon and Constantin Brancusi. They do not form a "school", each interpreting in his own way the suggestions of Cubist painting. Brancusi stands apart as a great individual whose aim it was to simplify in such a way as to get to the very essence of the object represented. "When you see a fish," he once said, "you think of its speed, its flashing, floating body seen through water. Well, I've tried to express just that." The beautifully polished flashing form of his "The Fish", as in the Museum of Fine Arts, Boston, is a masterpiece. Impressive in a similar way are his various conceptions of a bird in space (Plate 60). Zadkine (Plate 62) has used the combination and contrast of concave and convex forms with brilliant result in figures emblematic of poetry and music. Two great monuments, "The Destroyed City" at Rotterdam, commemorating the city's ordeal in the Second World War, and the monument to Van Gogh at Auvers-sur-Oise (1956), show how much of intensely human feeling it was possible to combine with modern technique.

The second of the century's adventures in expression, the Futurist movement, produced one remarkable sculptor in Boccioni. Like the Futurist painters he was concerned with the energy of movement and how it could be rendered in sculpture. It was his theory that "Objects never finish; they intersect with innumerable combinations of sympathy and innumerable shocks of aversion". Thus a bottle suggested infinite expansions of spirals and contrasts of angles and curves combined together. The idea of finding an equivalent for dynamic energy rather than representing a single phase of movement in an imitative fashion has a parallel in one of the works of the short-lived sculptor, Raymond Duchamp-Villon, "The Horse" (1914).

These were early attempts to convey an awareness of a new destiny in the twentieth century and, in the sense that they sought to define forces operating on the world, they may to some extent be likened to the efforts of the sculptor in primitive societies to define the forces of sun, rain and fertility. There have been other sculptors, however, who have looked back to the calm and serenity of the classical ideal. In France, Antoine Bourdelle produced very beautiful works inspired by an enthusiasm for ancient Greece. With him may be placed Aristide Maillol, whose figures of women recall those of Renoir (Plate 54), and Charles Despiau. Something of a similar static and monumental character is found in England in the work of Frank Dobson and Eric Gill. The German sculptor, Wilhelm Lehmbruck, though somewhat mannered in the elongation of his figures, shares a classic serenity with Bourdelle.

Broadly speaking, two roads were open to the sculptor's choice in the early twentieth century. There was on the one hand the traditional practice of representing the human being in a more or less lifelike fashion, for which the clay model to be cast in bronze, or a hard material directly carved, were the appropriate media. On the other there were the possibilities of form freed from imitation of this kind, "abstract" form with its own life using traditional materials in a new and flexible way. In favour of the first alternative it could be said that the real basis of sculpture is a human relationship. So eminent a modern sculptor as Henry Moore has said, "I don't think that we shall or ever should get far away from the thing that all sculpture is based on in the end, the human body".

Others, however, have seen in the twentieth-century developments of industry, science and technology a need for sculpture to participate in some way and to convey the spirit of an age in many ways so different from those that have gone before. For instance the American artist, Alexander Calder (trained, it is significant to notice, as an engineer), in developing his "mobiles", shapes of metal, joined and balanced in such a way as to move with a touch or breath of air, in changing relations, invites us to appreciate these as beautiful in themselves (Plate 61). It is clear that to give these metal forms some imitative semblance of humanity would be as much out of place as in the chassis of an automobile.

The crossroads of the century can be seen in comparing the early and later work of Jacob Epstein. For a while, between 1913 and 1916, he was influenced by the revolutionary suggestions of Cubism and the English movement, Vorticism, with the attachment of the latter to the idea of the machine. Brancusi, whom he met, inspired an enthusiasm for the abstract, simplified form which appears in the "Doves" of 1915. His friend, the brilliant sculptor Gaudier-Brzeska, encouraged him to look to primitive example. An ardour for machinery produced the "Rock Drill" (Tate Gallery) of 1913, a robot, "the armed, sinister figure of today and tomorrow", as Epstein himself later described it.

He soon turned away, however, from what he

called the "terrible Frankenstein's monster" from which humanity was absent, to his monumental groups and figures. They may be compared with those of Rodin as expressing a series of thoughts about the nature of man and woman in which there is a strong Biblical element. A remarkable gallery of portraits contains much of his best work. His "Madonna and Child" (Cavendish Square, London) (Plate 59) and "Lazarus" (New College, Oxford) are outstanding among his monumental sculptures in power and feeling, though not "modern" in any specific aspects of style.

Modern ideas led a number of artists in Europe in a completely opposite direction. Not humanity but space, the action of light on different substances, the interplay of geometric forms "beautiful in themselves"—as Plato long ago had imagined they might be—were aims which produced some remarkable results. Among them is the work of the Belgian sculptor, Georges Vantongerloo. A fugitive at The Hague in the war of 1914–18, he became associated with the Dutch group, de Stijl—artist philosophers who sought a language of form which could be applied equally to architecture, sculpture, painting and various forms of design. The rectangle was its basis and Vantongerloo produced a number of abstract sculptures consisting only of verticals and horizontals. From this he went on to the exploration of the idea of space in constructions of wire and plexiglass. In what way could these be described as beautiful? To explain the beauty he sought the artist has used the analogy of the aurora borealis. This phenomenon of nature, "the result of the actions of electrons in a magnetic field", was a new source of inspiration.

Transparent constructions open on all sides and allowing the light to penetrate through, creating varied effects of light and shade, have been the special product of the Russian "constructivists", Naum Gabo and Antoine Pevsner. Gabo has used stretched wires and plastic in intricate patterns of line, Pevsner intersecting plates of metal and plastic (Plate 64). Purely abstract form of a more monumental kind is to be found in the sculpture of Barbara Hepworth, though she sometimes uses the linear effect of wires attached to some solid form—almost suggesting some strange kind of harp. The considered relation of mass and line, of solids and scooped-out hollows, the refinement of proportion and of surface texture apart from any association with living organisms or natural objects are the aim. The result of this work is an exceptional purity of visual effect.

Barbara Hepworth preceded Henry Moore in developing the contrast and combination of an outer and inner form and would seem to have influenced his work in that respect, though the art of Moore has a much wider range. As he himself has said, it is the human figure that interests him most deeply, and this link with the tradition of sculpture is a part of his strength. He has profited also by a great receptiveness to powerful forms in the sculpture of the past—archaic Greek, Romanesque, Pre-Columbian. Ancient Maya and Mexican sculpture in particular impressed him, and the reclining rain-god (Chac-Mool) figure (Plate 15) was a source of inspiration for his many versions of figures in the same reclining pose (Plate 56).

He has also been much aware of the direction of thought in the art of his own time. The Surrealist movement of the nineteen-thirties, which drew attention to the "found object", that is to say forms in nature—pebbles, rocks, bones, tree roots, shells—possessing some unusual interest of shape appealing to the imagination, has influenced his outlook. The result may be seen in those reclining figures which one can if need be, looked at as a series of massive boulders taking human shape. In the fashion of other modern sculptors he has made much use of open construction—of hollowed out and concave elements as well as solids—yet essentially his art is one of mass and repose. More important in the result is the sense he gives of an "ideal" form—ideal in majesty and dignity. This is his greatness rather than any particular technical device, and it is in this that he stands out among contemporary sculptors and in relation with the giants of the past (Plate 58).

A different effect has been exerted on other sculptors by the imaginative or fanciful ideas of Surrealism, causing them to move away as far from a purely geometrical and abstract conception of the art as from academic imitations of the past. It has led some to favour the fantastic and grotesque or alternatively to convey some sense of hidden forces in nature. Thus the sculpture of Jean Arp, abstract in the sense of not imitating any specific form, is intended to suggest ideas of growth and rhythm. Much figure sculpture again, as in the distant past, suggests the mystery of the idol or uses various distortions to make some comment, sometimes bitter, pessimistic or satirical, on the condition of man. The strange figures, emaciated and elongated, of Alberto Giacometti, give a feeling of human isolation and suffering and, for all their impossible leanness, an intense reality (Plate 63).

The quest of strangeness has produced other remarkable results—the fantasies of Max Ernst, the phantom delicacies of Germaine Richier, the work of a number of the younger English sculptors in the nineteen-fifties, Kenneth Armitage, Lynn Chadwick, Bernard Meadows and others, whose hybrid and sinister forms might be taken as a product of the disturbed modern consciousness in the atomic age. The metals iron, steel and aluminium, have had a special fascination for a number of artists. Julio Gonzalez, the friend of Picasso, whom he instructed in ironwork, was a modern pioneer in the application of the blacksmith's craft, forging figures as arresting as the barbed "Cactus Man" (1939–40).

The forms of metal sculpture are various, though in general it may be said that working directly in the material entails a departure from the solid form with human semblance. It is distinct from a carving in marble, wood or stone or of the clay model cast in bronze (where the metal is only a means of reproduction). The workers in iron often produce a kind of linear effect, a metal silhouette, as in the early productions of the English sculptor, Reg Butler. If a comment on the human figure is made in iron it is likely to become as robot-like and caricatured as in the work of the Italian sculptor, Roberto Crippa. Yet an affection for the qualities of metal in itself and, by implication, for the machine forms of today, appears in the work of artists in many countries, in America with Calder, whose mobiles have already been mentioned, and such artists as David Smith and Herbert Ferber. Richard Lippold is of note for a remarkable abstract construction of gold wires "The Sun", commissioned by the Museum of Modern Art, New York, and dazzling in its fine-spun complexity. The German sculptor, Norbert Kricke, the Spanish sculptor, Eduardo Chillida, and the English sculptors, Kenneth Martin and Anthony Caro show this affection, either for the quality of the bar of iron or the twisting linear forms metal can be made to assume (Plate 57). In France, Berto Lardera has worked on a large scale with sheets of wrought iron that take on striking and graceful shapes. The clear-cut character of scaffolding and girder have inspired the Hungarian sculptor, Nicolas Schöffer.

Some find a fascination in the "ready-made" forms of iron, discarded machine parts and such things, using them to build up fantastic forms, like the French sculptor, César, who has made interesting objects out of scrap, and in England Eduardo Paolozzi. It is a further development of the mechanical bent of modern sculpture to add the effect of actual movement. Thus Jean Tinguely has devised "mobiles" which go beyond those of Calder in being set in motion by mechanical power, the effect being heightened by harsh mechanical sound.

The experimental attitude of a number of sculptors is illustrated by the assertion of Victor Vasarely, who has produced both painting and sculpture, that these are now "anachronistic terms: it is more exact to speak of a two-, three- and multi-dimensional plastic art". In one way or another a number of artists have disregarded the old division between painting and sculpture, using elements of both in the same work. Abstract screens producing changing effects of space and movement to the eye have been produced by Vasarely, the South American artist, J. R. Soto and others. Another development seems to stem from the work of Kurt Schwitters who used all kinds of odds and ends, in painted reliefs. The varieties of surface texture and the curiously vivid qualities detail thus obtained have been much exploited.

Viewed as a whole, twentieth-century sculpture is a spectacle of great activity. The age has become surprisingly fertile in new forms, a forest of new ideas and technical departures. It is quite distinct from other periods if one concentrates on the *avant-garde* aspects and not on the work of a more conventional and imitative kind. Some people may lament the fact that the ideal beauty of the figure which was once the theme of the sculptor has lost its place, that the age is not in that sense "humanist" like the Periclean age of ancient Greece or the Italian Renaissance. Yet it still makes its effort to define and give visible shape to the forces that operate on a period of scientific adventure, a period in which new conceptions of the universe and of man's place in it are born. In that way it is related to all those epochs of past sculpture in which the ideas and progress of humanity are on permanent record. It is linked with what has gone before not only as a handicraft with its own rules but also as the crystallization of thought and, as a modern sculptor has said, the transformation "of an immaterial vision into matter".

CHAPTER XI

SCULPTURE AND THE PUBLIC

It may be asked if there are any simple rules to guide the observer in the appreciation of sculpture. Evidently one has to approach it rather differently from a picture which gives the illusion of three dimensions of sky or distance or the relation of objects on a plane surface. There is no such illusion in sculpture, except in so far as there is a resemblance to pictorial composition in some forms of relief. In the round it has a real existence and may be appreciated from this point of view, in a material way. By making a circuit round a work, one can admire the changes of form when seen from different viewpoints. Degas when showing favoured intimates his statuettes, is said to have turned them slowly round, lighting them at the same time in such a way that the shadows thrown on the wall brought out the variations of shape with added emphasis. The moving shadow thrown has likewise been observed as part of the fascinating effect of the "mobile".

To some extent it is a test of good sculpture that it "composes well" when seen from any angle, which means that it has that completeness of conception on which Michelangelo laid stress. Another thing to look for is the quality of the material used and the way in which the sculptor has brought it out. The variations of texture and colour are many, from the granular surface of a hard stone to the smoothness of ivory, from the milky sheen of alabaster to the delicate lustre of jade. There can be a pleasure in what is near to being a physical sensation. Without actually touching, one may be keenly aware of the "tactile" quality. Here the sculptor makes use of the suggestions towards art that nature has already provided. This is one of the reasons why sculpture looks best in the open air. Its link with materials in nature then comes into view and even the weathering of statues which have been long exposed to the elements has a satisfaction to give like that which we derive from seeing how pebbles have been worn by the sea, or rocks by the action of climate. Exhibitions of sculpture in recent times, in such natural settings as public parks provide, have been a reminder of the pleasure afforded to the great connoisseurs of the past, the Medici of Florence for example, by the statuary in the gardens of their palaces. In its circumscribed modern form the garden is still the most apt of places for sculpture, a fact subconsciously realized by those who decorate it with plaster rabbits and gnomes, although these are sadly far from being works of art and a poor substitute for sculpture in the true sense of the word.

Again, we may look for and find pleasure in the order a sculptor imposes on his material. The ordered rhythm, the system of folds with which the Greek sculptor made drapery so suggestive of graceful movement is an instance. Rodin regarded his art as one of light and shade and the intended effect which the cast or carving gained by the polished highlight or the shadow produced by a deep concavity: a prime source of effect or, indeed, of many effects varying with the light in which the work of sculpture is seen. How different these can be becomes visible in the photographs of sculpture. They vary as much as the reproductions of a musical composition, and a photograph will often reveal a force or a subtlety of carving or modelling that might otherwise have gone without notice, though the capability of producing this effect was nevertheless inherent in the original work.

Sculptors using metal convey as much pleasure in the qualities of material as those using stone or wood. Corrosions of surface often have a special interest for the eye, an instance being the green patina which bronze takes on. On the other hand a highly polished surface, catching the sparkle of light and giving a reflection of surrounding objects like a mirror, has an added physical attraction. The polished bronzes of birds in space and fish by Brancusi are remarkable for the effect thus gained.

The association of sculpture with movement makes use of one of the primary and universal means of catching the eye. It accounts for the particular attraction of fountain figures which are the vehicle of running water. Though we regard the "mobile" as a modern invention even this has its historical counterpart. Thus the early sixteenth-century Florentine sculptor, Rustici, devised a fountain figure of Mercury standing on a globe and holding a mechanical device described as having "four delicate wings like those of a butterfly". They were rotated by water spouting from the figure's mouth.

Logically speaking, there is no absolute difference between moving round a work of sculpture to appreciate its variety of form from different angles and standing still to observe a sculptured object which itself moves. It is true that many statues depend for their effect on the static qualities of weight and repose, but we must allow for the difference of aim in such works as Calder's thin sheets of metal and find a separate pleasure in their changing relations as they move.

In all, one can point to many material qualities in sculpture capable of giving satisfaction, not excluding the "kinetic" and "optical" experiments of the present-day; to aspects of the art in terms of mass, line and proportion which may be considered "beautiful" in a purely aesthetic sense. Yet there remains something mysterious in sculpture and this, though it cannot be simply explained, is one of its great assets. In many magnificent works it is clear that the aim was not that of providing an exact imitation of objects in nature. Indeed if this were the criterion we should have to regard the effigies of celebrities in a waxworks as masterpieces (which they may be but in a completely different category).

Nor can it be said of many master works that their creators were devoted to a purely artistic ideal or that their aim was solely self-expression. The general impression that may be derived from an historical survey, even in outline as in the foregoing chapters, is that sculptors in the great periods have not worked in isolation but in a collective fashion, which is to say that personal expression was subordinate to other aims. The Gothic mason or "imager" must certainly have delighted in the exercise of his craft, but it was not his function to create artistic effects for the admiration of trained connoisseurs. His taste was to provide an image of belief for the simple and unlettered. If he allowed play to his fancy in detail, it was within the framework of the collective purpose.

What the sculptor expresses is the spirit of his age and, even more than that, aspirations which go beyond the material circumstances of any time. Self-expression belongs to a freer art such as painting. The more difficult materials in which the sculptor works make for another outlook. Even so great an individualist as Michelangelo indicates, in that famous remark earlier quoted, that a sculptor is a medium, as it were, in bringing a work into existence and not merely one who imposes his ego. It is the triumph of the spirit in his conflict with stubborn solids that calls for appreciation of a more emotional or deeper kind than the pleasure produced by the grain of wood or the crystalline quality of a white marble.

The word "spirit" which one is bound to use stands for that mysterious element which is none the less present and discernible in great sculpture for being hard and perhaps impossible to put into words. It enters even into portraiture. What we find moving in Roman portrait-sculpture is not only the fact that it provides us with a description of the features of historical personages. One might go so far as to say that the sculpture somehow penetrates to the very soul of these beings. There is the essence of humanity in sculpture, as well as the record of historical progress, all of humanity's approach to the universe. At the present time, in which there has been so much new life in sculpture, this still holds good. For a section of the public it has the excitement of novelty. There are some who find its products eccentric. Yet the best of it is as significant of the hopes and fears, the problems and the aspirations of the world as any in the past. From this viewpoint, sculpture is a unity entailing no need to exclude from appreciation any one age, ancient or modern, more than another. All of them can help us to understand our own time.

49. ANDREA RICCIO (*above*): *Pan. Bronze.*
(*Oxford, Ashmolean Museum*)
ANTOINE BARYE: *Lion and Serpent. Bronze.*
1832–5

50. GIOVANNI BERNINI: *The Ecstasy of St Theresa, 1644–7*
(*Rome, S. Maria della Vittoria*)

51. FRANÇOIS ROUBILIAC: *The Nightingale Monument, 1761, Westminster Abbey*

52. JEAN-ANTOINE HOUDON (*left*): Bust of Voltaire, 1781

(*Victoria and Albert Museum*)

GIOVANNI DA BOLOGNA (*right*) *The Rape of the Sabines. Bronze.*

(*Victoria and Albert Museum*)

53. GRACE IN FRENCH SCULPTURE
(*above*): EDGAR DEGAS:
La Grande Arabesque. Bronze. c. 1885–90 (*Tate Gallery*)
(*below*): JEAN GOUJON: *Nymphs from La Fontaine des Innocents. Sculpture of* 1547–9 (*Louvre*)

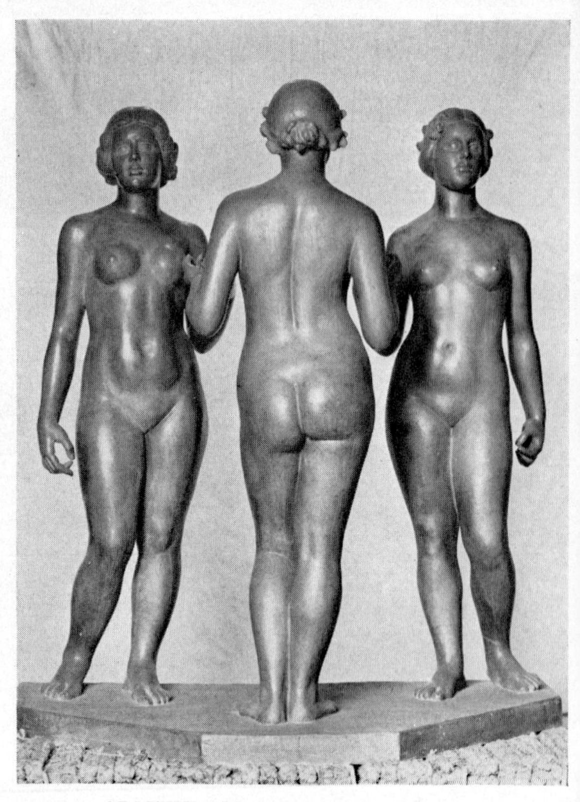

54. ARISTIDE MAILLOL: *The Three Nymphs. Bronze.* 1930–7
(*Tate Gallery*)

55. AUGUSTE RODIN: *Balzac*, *1898*
(*Musée Rodin, Paris*)

56. HENRY MOORE: *Reclining Figure, Three Piece: Bridge Prop, 1963*

57. ANTHONY CARO: *Early One Morning.* *Contemporary iron sculpture in open-air setting*

58. HENRY MOORE: *King and Queen* (*in situ* in Scotland). *Bronze*. *1952–3*

59. JACOB EPSTEIN: *Madonna and Child. Bronze. 1952*

(Convent of the Holy Child, Cavendish Square, London)

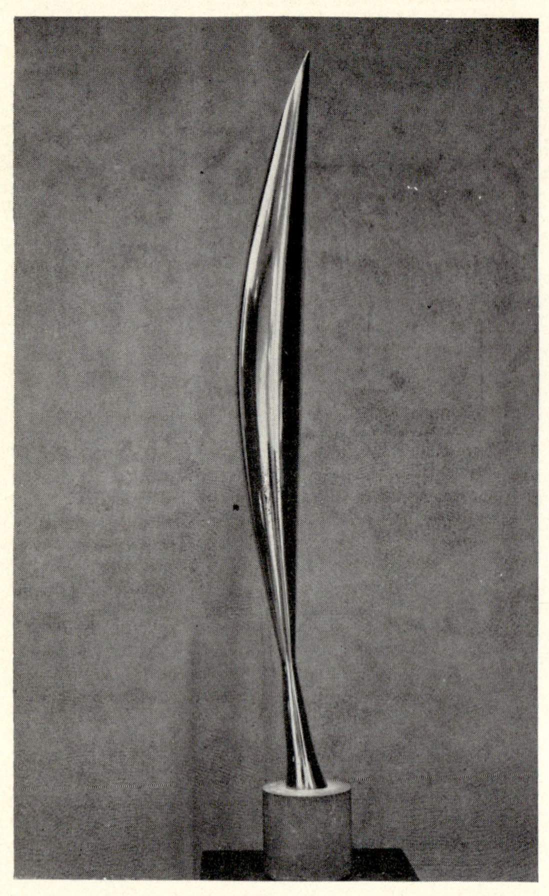

60. CONSTANTIN BRANCUSI: *Bird in Space.*
Bronze. 1919
(*Museum of Modern Art, New York*)

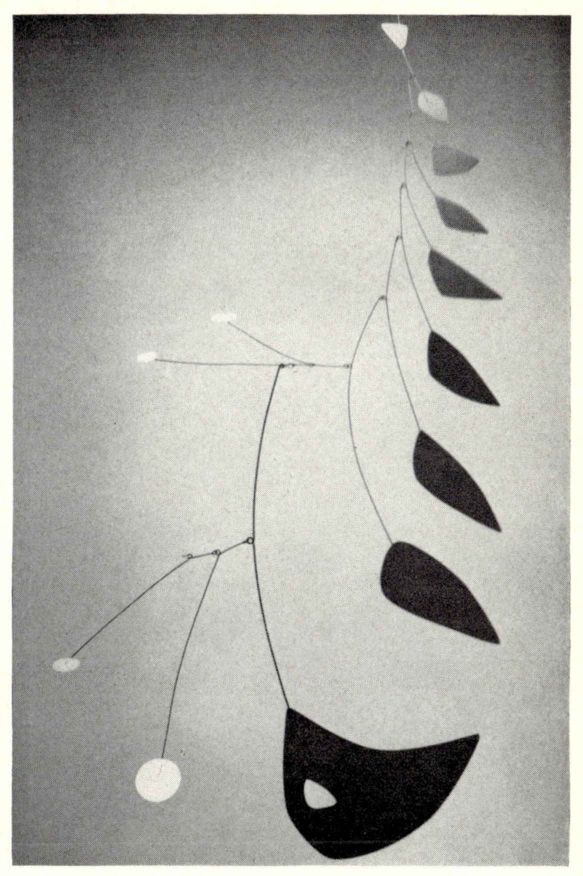

61. ALEXANDER CALDER: *Mobile: Four White Dots, 1959*

62. OSSIP ZADKINE: *Birth of Forms. Plaster. 1947*

63. ALBERTO GIACOMETTI: *Walking Man.*
Bronze. 1950

64. MODERN SCULPTURE IN RELATION WITH ARCHITECTURE.

ANTOINE PEVSNER: *Bird in Flight*

(*General Motors Building, Detroit*)

A GLOSSARY OF TERMS

Acanthus. Plant serving as the motif of the carved decoration of a Corinthian capital.

Abstract. In sculpture, refers to form which represents some ideal or visionary conception rather than the reproduction of an object in an imitative fashion.

Baroque. Term applied to architecture, sculpture and painting, in Europe, in the sixteenth and seventeenth centuries, in sculpture applying to work of an extravagant and emotional character.

Baldacchino. A canopy supported on columns.

Boss. A round prominence at the intersections of vaulting often carved with ornament in Gothic buildings.

Capital. The head of a column, lending itself to a variety of sculptured ornament, e.g. in Byzantine, Gothic and Classical architecture.

Caryatid. Human figure doing duty as a column, as, e.g. in the Erechtheion at Athens.

Cire Perdue. Method of hollow casting by which a layer of wax modelled on a solid core is reproduced in bronze.

Classical. Term applied to the great period of Greek sculpture, fifth–fourth century B.C., and to work of other periods with a likeness of aim.

Classicism. Attempted revival of the classical past,

as in the neo-classic sculpture revival at the beginning of the nineteenth century.

Constructivism. Art movement originating in Russia in the early years of the Bolshevik Revolution, a new departure in sculpture in the construction of three-dimensional objects in metal, wire and glass, suggested by modern technology.

Cubism. New treatment of form and space in painting which influenced a number of sculptors to attempt equivalent methods of treatment from about 1910 onwards.

Found Object. Term applied to what were once regarded as curiosities or freaks of nature in which artists found new interest of form such as has had some influence on twentieth-century sculpture.

Frieze. The middle part of the entablature above the columns of a classical building providing a continuous surface for sculpture.

Gargoyle. Water spout carved in grotesque animal or human form by Gothic sculptors.

Gothic. Comprises the sculpture produced in Europe from the twelfth to the fifteenth century.

Metope. Spaces between the grooved triglyphs in the classical frieze of the Doric order, often carved in relief.

Misericord. Bracket on the underside of a choir seat. Often carved with grotesques by Gothic sculptors.

Mobile. Forms of metal sculpture so constructed as to move in various balanced relations.

Patina. Green surface tone produced by the action of time and oxidation on bronze sculpture.

Pediment. Triangular gable over the portico of a classical building. Often used for a sculptural composition.

Relief. Sculpture on a flat surface, sometimes in the form of low-relief, being almost two dimensional though high-relief may be nearly round in effect.

Renaissance. The great development of art centred in Italy in the fifteenth and sixteenth centuries in which sculpture had an important part.

Rococo. Ornate style in architecture, sculpture and painting which flourished in the eighteenth century.

Romanesque. Sculptural style of the eleventh-twelfth centuries in Europe, more especially connected with religious architecture.

BIOGRAPHICAL NOTES

Algardi, Alessandro (1602–54)
Italian sculptor who worked in Bologna and Rome, where he was the rival and contemporary of Bernini. He is noted for portrait sculpture such as his bronze and porphyry bust of Pope Innocent X (Palazzo Doria, Rome) and for superbly finished bronzes of religious subjects.

Antico (Pier Jacopo Alari Bonacolsi) (*c.* 1460–1528)
Italian sculptor of the Mantuan School who worked for the Gonzaga court, producing a number of fine bronzes inspired by antique models. A celebrated work is the Martelli Mirror with decoration in relief (Victoria and Albert Museum).

Archipenko, Alexander (1887–1964)
Russian-born sculptor, of note as the first to apply the ideas of Cubist painting to sculpture when working in Paris. Characteristics are elongated figures with a strong geometric element, sometimes painted, and using hollow spaces as well as mass to convey construction. He became an American citizen in 1928. Later experimental work included transparent sculpture lit from within.

Armitage, Kenneth (b. 1916)
English sculptor who uses a dramatic simplification in idol-like figures with rudimentary heads and limbs. His work has been considered by some critics

as an indirect comment on the problems and frustrations of modern life.

Arp, Jean (b. 1887)
Sculptor and painter of the modern French School. Associated with the abstract and surrealist movements, he developed an abstract sculpture intended to suggest natural forms and rhythms in nature without direct imitation of objects. His works include a monumental sculpture for the University Centre, Caracas, and a large bronze relief for Unesco, Paris.

Barlach, Ernst (1870–1938)
German sculptor and ceramist, noted for carvings in which he recaptured something of the spirit of medieval art. His work was proscribed during the Hitler régime in Germany.

Barye, Antoine Louis (1795–1875)
French sculptor and painter, the son of a goldsmith, famous for his bronzes and watercolours of wild animals, rendering their movement and ferocity with great power in such works as his "Tiger devouring a Gavial" (1831) and "Lion and Serpent" (Plate 49). He exemplifies the Romantic spirit in sculpture and had a number of followers in animal themes.

Bernini, Giovanni Lorenzo (1598–1680)
Italian architect and sculptor, the great exponent of the Baroque style. He worked mainly in Rome, completing the exterior aspect of St Peter's with its colonnaded piazza and the interior with the elaborate canopy over the high altar and many grandiose

monuments. The Fontana di Trevi, Rome, is a masterpiece of his fountain design. His principal works include "Apollo and Daphne" (Borghese Gallery) and "The Ecstasy of St Teresa" (church of S. Maria della Vittoria, Rome) (Plate 50) and numerous portraits.

Berruguete, Alonzo (1480–1561)
Spanish architect, sculptor and painter who studied under Michelangelo and worked in Spain in a style influenced by that of the Italian Renaissance.

Bertoldo (*c.* 1420–91)
Italian sculptor, one of the principal pupils of Donatello whom he followed in style. He worked for Lorenzo the Magnificent at Florence, one notable production being a battle relief after the antique (Museo Nazionale, Florence).

Bird, Francis (1667–1731)
English sculptor, who worked for Sir Christopher Wren. Reliefs in the portico of St Paul's and the statues of the apostles of the west front are by him.

Boccioni, Umberto (1882–1916)
Italian sculptor and painter, associated with the Futurist movement. In 1912 he drafted a "Technical Manifesto of Futurist Sculpture" and produced some original works inspired by the aim of representing dynamic energy.

Bologna, Giovanni da (1524–1608)
French architect and sculptor who worked in Italy. An outstanding sixteenth-century master, he is noted for works which show some of the exaggerations of Mannerism. Principal sculptures are the groups for the Fountain of Neptune, Bologna, his bronze "Mercury" and "The Rape of the Sabines" (Plate 52).

Bourdelle, Antoine (1861–1929)
French sculptor in the classic manner. He studied with Dalou and was for a long time a collaborator with Rodin, though a sympathy with Greek art is more marked in his principal productions. His "Herakles" and reliefs for the Théatre des Champs-Elysées, Paris, are celebrated works. There is a Bourdelle Museum in Paris.

Brancusi, Constantin (1876–1957)
Rumanian sculptor, one of the great masters of modern times. Individual in style and aim, he succeeded in expressing the essential character of natural forms by a reduction to an extreme simplicity. Versions of "Bird in Space" (Plate 60) and "The Fish" are celebrated examples. An impressive monument in Rumania, at Turgu-Jiu near his birthplace, is his "Endless Column". He left his studio and its contents in Paris as a Brancusi Museum.

Butler, Reg (b. 1913)
English sculptor trained as an architect, who worked as an ironsmith during the Second World War and took up sculpture in 1944. The linear effect of an

iron framework distinguishes his early productions though he later turned to a more complete rendering of the human figure.

Calder, Alexander (b. 1898)
American sculptor, noted for the invention of metal constructions capable of being set in motion ("mobiles"). After studying engineering he made his first animated objects and sculpture in wire in 1926. His mobiles were first shown in 1932. He has been called a "draughtsman in space" and his work has had considerable influence.

Cano, Alonzo (1601–67)
Spanish architect, sculptor and painter who was known as the "Michelangelo of Spain". He worked at Seville, Madrid and Granada, where there are typical examples of his work in the cathedral.

Canova, Antonio (1757–1822)
Italian sculptor, noted for his efforts to create a purified form of classicism in sculpture which made him the central figure of the "Neo-Classic" revival in Rome, though often considered cold and artificial. A life-size nude statue of Napoleon I as an antique hero, with a winged statuette of Victory in his hand (Brera Gallery, Milan) illustrates his ideas.

Caro, Anthony (b. 1924)
English sculptor, trained in engineering who turned to sculpture in 1947, concentrating on large metal forms and then balance and relation in space.

Carpeaux, Jean Baptiste (1827–75)
French sculptor, who has an important place in French nineteenth-century art as one who continued

the elegance of past tradition and was also a forerunner of the greater freedom of Rodin. His group of "La Danse", executed for the Paris Opéra in 1869 was one of his principal works, noted for its vigour.

Cellini, Benvenuto (1500–71)
Italian sculptor and goldsmith who produced one of the Renaissance masterpieces in his bronze "Perseus with the Head of Medusa" (Loggia dei Lanzi, Florence) (Plate 48). A famous work of craftsmanship is the salt-cellar wrought in solid gold for Francis I (Vienna). His Autobiography not only includes much exciting adventure but is also highly informative about his technique as a sculptor—as in his account of the casting of the "Perseus".

César (César Baldaccini) (b. 1921)
French sculptor who uses iron and miscellanies of scrap metal material to build up impressive and mysterious forms.

Chadwick, Lynn (b. 1914)
English sculptor who first produced mobiles and abstract constructions in the nineteen-forties but has since gained international repute for figurative work of an imaginative character.

Chillida, Eduardo (b. 1924)
Spanish sculptor, first trained as an architect, noted for abstract sculpture in iron. He executed a monument to Sir Alexander Fleming in San Sebastian in 1955.

Cosmati, The (twelfth–thirteenth century)
Roman family of architects, sculptors and mosaic workers whose work for ecclesiastical buildings had a wide influence, as in the shrine of Edward the Confessor at Westminster Abbey.

Coustou, Guillaume (1678–1746)
French sculptor, pupil of Coysevox, together with his brother Nicolas (1658–1733). He continued the monumental tradition in such works as the horses of the entrance to the Avenue des Champs-Elysées, Paris.

Coysevox, Antoine (1640–1720)
French sculptor who worked for Louis XIV and gave to his sculpture the impressive authority of the Baroque style. The tombs of Mazarin and Colbert in St Eustache, Paris, statues in Paris and at Marly and Versailles, and portraits such as his fine bust of Condé (Louvre) are principal works.

Dalou, Jules (1838–1902)
French sculptor who studied under Carpeaux. A refugee in London in 1871, he became a professor at South Kensington and had some influence on English sculpture, though returning to France in 1879. "The Triumph of the Republic" (1900), Place de la Nation, Paris, was one of his principal monumental works.

Daumier, Honoré (1808–79)
Though best known as caricaturist, the French

artist, Daumier excelled both as painter and sculptor. The figures he modelled to provide suggestions for his lithographs, little known in his own time, are rich in character and intensely vivid in conception, the raffish character "Ratapoil" (1849) being one memorable creation.

Degas, Edgar (1834–1917)
Mainly known as painter and draughtsman, the great French artist Degas was always inclined towards sculpture, his earliest works dating from 1866, though it was late in life that he produced the studies of dancers in motion, and racehorses, now so highly esteemed. Many wax figures were cast in bronze after his death. Unusual in realism was his "Dancing Girl of Fourteen" with real hair and dress, exhibited in 1881, a version being in the Tate Gallery.

Della Quercia, Jacopo (1374–1438)
Italian sculptor, the son of a goldsmith, a precursor of Michelangelo in breadth of style. Statues for the Fonte Gaia, Siena (1409–19) and reliefs for the church of S. Petronio, Bologna, are principal works.

Della Robbia, Luca (1399–1482)
Italian sculptor, noted for the refinement of feminine type and beauty of execution in bas-reliefs and the enamelled terracotta known as "Robbia ware". His masterpiece was the series of reliefs for the Cantoria of the Cathedral, Florence. His nephew Andrea (1435–1525) and Andrea's sons continued the production of ceramic sculpture.

Despiau, Charles (1874–1946)
French sculptor who represents a modern style of

restrained classicism. He worked for Rodin from 1907 to 1914 and produced many fine portrait busts and nudes.

Dobson, Frank (1889–1963)
English sculptor noted for single figures nude or draped, "Truth" (Tate Gallery) being an example on a large scale, and for portrait busts including the "Osbert Sitwell" in polished brass and "Lydia Lopokova" (Tate Gallery). He also produced many small terracottas.

Donatello (*c.* 1386–1466)
Italian sculptor, one of the greatest representatives of the Renaissance in Florence. He worked in Siena, Florence and Padua and had many followers in work in bronze. Masterpieces by him are the great equestrian statue of Gattamelata (Padua) the bronze "David" (Bargello, Florence) and his "Judith and Holophernes" (Loggia dei Lanzi, Florence) and his "St George" (Plate 42). He made happy renderings of childhood as in figures for the font at the Baptistery, Siena, but showed also a unique capacity for dramatic and powerful expression.

Duchamp-Villon, Raymond (1876–1918)
French sculptor, first influenced by Rodin but later associated with the Cubist movement. Among his works are the "Head of Baudelaire" (1911) and the dynamic "Horse's Head" (1914), the latter of especial note in the evolution of modern sculpture.

Epstein, Sir Jacob (1880–1959)
American-born sculptor who settled in England,

becoming one of the most prolific, powerful and controversial of English sculptors. The "Rock Drill" (1913) (Tate Gallery) represents a brief excursion into "modernism"; but controversy was aroused mainly by his emotional emphasis. His work comprises architectural sculpture, from the figures on the British Medical Association building (1907) (now vanished), to the "Christ in Majesty" for Llandaff Cathedral (1957); memorials (e.g. "Rima" in Hyde Park); symbolic single figures and a remarkable series of portrait busts cast in bronze.

Ernst, Max (b. 1891)
German-born sculptor and painter, one of the founders of the Surrealist movement. Perhaps best known for imaginative paintings and collages, he shows a kindred sense of the fantastic in sculpture, in works such as the "King and Queen" (1944), "The Parisian Woman" (1950), and "Oedipus" (cast in bronze in 1960).

Flaxman, John (1755–1826)
English sculptor who is numbered with Canova and Thorwaldsen as one of the leaders of Neo-Classic revival, his outline illustrations to Homer being as influential in this respect as his sculpture. He designed many monuments (including those to Nelson and Reynolds in St Paul's Cathedral). His classical designs for Wedgwood china are celebrated.

Gaudier-Brzeska, Henri (1891–1915)
French-born sculptor who worked in England. Res-

ponsive to the forward ideas of the time (Cubism, Vorticism), he produced a few works of great brilliance and force before he was killed on active service in 1915.

Gauguin, Paul (1848–1903)
The great French Post-Impressionist painter was early in the appreciation of primitive art which has been a feature of the twentieth century, and his own carvings in wood, inspired by his stay in Polynesia, though few in number, are important in this respect.

Ghiberti, Lorenzo (1378–1455)
Italian sculptor, famous for one of the early masterpieces of the Renaissance, the bronze gates of the Baptistery, Florence, declared by Michelangelo to be worthy of the gates of Paradise. Reliefs of scenes from the Old Testament appeared on the first gate (1424) and from the New Testament on the second, completed in 1452. He produced other statues, reliefs and tomb sculpture in Florence and Siena.

Giacometti, Alberto (b. 1901)
Swiss-born sculptor and painter, noted for modelled figures of unusual proportion, thin and elongated, which have evoked ideas of the isolation and suffering of humanity. They are the later stage of an evolution which began with works of a cubist and abstract nature and turned to the fantastic as in the "Palace at four in the Morning" (1932–33). "The Pointing Man" (Tate Gallery) is a typical work (1947).

Gibbons, Grinling (1648–1721)
English sculptor and wood-carver, best known for the decorative still-life carvings with which he adorned many buildings, religious and secular, being especially associated with Sir Christopher Wren. In his choir stalls and other work at St Paul's and in mansions such as Blenheim and Petworth he showed himself both artist and master-craftsman. He executed statues also (Charles II, James II) but these are less outstanding than his sculptural decoration.

Gilbert, Sir Alfred (1854–1934)
Sculptor and goldsmith who worked mainly in metal, best known for the aluminium statue of Eros on the Shaftesbury Memorial Fountain at Piccadilly Circus (1893). Other works were the tomb of the Duke of Clarence at Windsor and the Queen Alexandra Memorial in London (1932).

Gill, Eric (1882–1940)
Sculptor, engraver and type-designer, noted for his revival in England of stone carving and carved lettering. His principal works include the Stations of the Cross, Westminster Cathedral (1913–18), and the exterior sculpture on Broadcasting House, London (1929–31).

Girardon, François (1628–1715)
French sculptor, one of the artists who worked for Louis XIV under the direction of Charles Le Brun, though his work is less formalized and more sensuous than that of his court contemporaries, the relief of "The Bath of the Nymphs" at Versailles being an example.

Gonzalez, Julio (1876–1942)
Spanish-born sculptor, son of a jeweller. He settled in Paris and took to iron sculpture about 1927. He was a pioneer in this medium, his work displaying both charm and wit. Considered his masterpiece is the life-size figure of a woman, "La Montserrat" (1937), inspired by the Spanish civil war and now in the Stedelijk Museum, Amsterdam.

Goujon, Jean (*c.* 1515–68)
French sculptor, an outstanding figure of the Renaissance period in France, his work being distinct in its flowing rhythms and gay and sensuous charm. One of his principal achievements was the series of reliefs for the Fontaine des Innocents, Paris, executed in 1549.

Hepworth, Barbara (b. 1903)
English sculptor, noted for her abstract carvings. She studied direct marble-cutting in Italy and was influenced in her work by Brancusi and Arp, though developing a distinct style which has gained international appreciation.

Houdon, Jean Antoine (1741–1828)
French sculptor who excelled in portraiture. He visited America in 1785 to execute a statue of Washington and produced many busts of famous contemporaries, including those of Voltaire, Rousseau, Lafayette, Mirabeau and Napoleon.

Kricke, Norbert (b. 1922)
German sculptor, an aviator in the Second World War, who later came into prominence with work in steel wires, showing a striking sense of design.

Laurens, Henri (1885–1954)
French sculptor, inspired by the Cubist movement to produce reliefs and polychrome works of a geometrical kind. His later work became more rounded and poetic in theme but still seeking first essentials of form.

Lehmbruck, Wilhelm (1881–1919)
German sculptor, noted for graceful figures of classic simplicity. Examples are in the Museum of Modern Art, New York and the Museum of Duisberg, near which city he was born.

Leonardo da Vinci (1452–1519)
The sculpture of the great Leonardo, like so much of his effort, remains a magnificent idea rather than a material product. His equestrian monument to Lodovico Sforza at Milan got no farther than the model stage. Three bronzes of horses exist, however, by or after Leonardo, probably made from his wax models for the projected fresco, the Battle of Anghiari. One is at Budapest, a second in the Metropolitan Museum, New York, the third in an English private collection.

Le Sueur, Hubert (*c.* 1595–1650)
French sculptor who worked in England, his equestrian statue of Charles I in Trafalgar Square being considered one of the best pieces of open-air sculpture in London. The Lennox and Richmond monument in Westminster Abbey and royal statues at St John's College, Oxford, are other works by him.

Lipchitz, Jacques (b. 1891)
Lithuanian-born sculptor who worked in Paris and was one of those who applied the geometric construction of Cubism to sculpture. He later worked in the United States.

Lippold, Richard (b. 1915)
American sculptor noted for constructions in metal wire, among them "The Sun" (1953–56), for the Museum of Modern Art, New York, made of gilt wires, an elaborate work involving 14,000 weldings.

Lysippus of Sicyon (*c.* 336–270 B.C.)
Greek sculptor, regarded as the great realist of his time, and the inventor of a system of proportion, which gave a new grace to length of limb in proportion to head. He was noted also for portraiture. The athlete scraping himself (Apoxyomenos) in the Vatican is one of the copies by which his work is known.

Maillol, Aristide (1861–1944)
French sculptor and graphic artist, the great classicist of modern times. A student of Greek sculpture he reflects its serene dignity in his figures, though he added a directness and simplicity of his own. Monumental works include a First World War memorial at Banyuls, his native place, and "La Douleur" at Céret.

Manzù, Giacomo (b. 1908)
Italian sculptor who works in the Renaissance tradition, influenced by Donatello.

Marini, Marino (b. 1901)
Italian sculptor noted for his series of horses and riders in which he has specialized since 1936. Greatly simplified, they retain the feeling of life.

Martin, Kenneth (b. 1905)
English sculptor who has specialized in abstract metal constructions ingeniously relating form and motion.

Matisse, Henri (1869–1956)
The eminent French painter produced a number of sculptures between *c.* 1900 and *c.* 1930, showing much plastic freedom. A series of bronzes of the back view of a nude figure, produced at various dates (in the Tate Gallery), illustrates his approach.

Meštrović, Ivan (1883–1961)
Yugoslav sculptor who led a nationalist movement in art, designing the Serb national temple at Kossovo. From 1947 he worked in the United States. In addition to monumental works and statuary he produced a number of portrait busts including those of President Hoover and Sir Thomas Beecham.

Michelangelo Buonarroti (1475–1564)
The greatest of sculptors, as well as architect, painter and poet, Michelangelo first studied the antique under Bertoldo, Donatello's pupil, his early work showing the influence of ancient Roman reliefs, Donatello and Jacopo della Quercia, preluding the achievements of matchless individuality and power. Famous productions in order of date are the early "Battle of the Centaurs" (Casa Buonarroti,

Florence), the statue of "Bacchus" (1496–97) (Bargello, Florence), the life-size "David" (1504) (Accademia, Florence), the "Taddei Madonna" (1505–6) (Royal Academy) (Plate 44), the figures of "slaves" for the tomb of Julius II (*c.* 1513), the "Moses", (*c.* 1516) of the completed tomb (S. Pietro in Vincoli, Rome), the figures of Dusk and Dawn, Night and Day (1524–31), for the Medici tombs (S. Lorenzo, Florence) (Plate 43), and the "Rondanini Pietà", his last work (Castello Sforzesco, Milan) (Plate 45). He expressed every aspect of feeling and emotion through the human figure alone.

Mino da Fiesole (1431–84)
Italian sculptor whose work is noted for its finish and delicacy. He worked in Florence, Prato and Rome and produced a number of portrait busts and profile portrait reliefs as well as tomb sculpture.

Modigliani, Amedeo (1884–1920)
Italian-born painter and sculptor, whose few sculptures show great originality. He was inspired by Brancusi and also adapted the emphatic contours of Afrjcan sculpture, without loss, however, of European character.

Moholy-Nagy, Laszlo (1895–1946)
Hungarian sculptor and experimentalist, noted for abstract sculptures and reliefs in glass, metal and plastic.

Moore, Henry (b. 1898)
English sculptor, unequalled among modern sculptors in monumental power. A wide study of natural

forms, of pre-Columbian and archaic sculpture, as well as the influence of Picasso and the Surrealists, were the basis of a great individual development. His monumental works include the "Three Standing Figures" (Battersea Park, London), "Madonna and Child" (St Matthew's Church, Northampton), "Family Group" (Stevenage New Town), "Warrior with Shield" (Arnhem), "Reclining Figure" (Unesco, Paris). Versions of his favourite theme, the reclining figure, are widely distributed in open-air settings (Plates 56, 58).

Myron (fifth century B.C.)
Greek sculptor noted as a master of athletic movement, as in the famous discus thrower "Discobolos" preserved at Rome (Plate 26). Copies of a bronze group of Athena and Marsyas (*c.* 450 B.C.), once on the Acropolis, show other aspects of this capacity to seize a momentary attitude.

Nollekens, Joseph (1737–1823)
English sculptor, the author of many portrait busts, including those of Garrick, Sterne and Dr Johnson.

Paolozzi, Eduardo (b. 1924)
English sculptor born of Italian parents, who has made imaginative use of machine-like forms as in his "Artifical Sun" (1964).

Pevsner, Antoine (b. 1884)
Russian sculptor who, together with his brother, who took the name of Naum Gabo (b. 1890), pursued the experimental development known as constructivism. Both worked on a small scale in constructions of wire and blades of metal but also produced monumental abstract works, Pevsner for

General Motors (Plate 64), Detroit, 1955, and Gabo a remarkable monument in steel at the Bijenkorf Stores, Rotterdam.

Phidias, (*c.* 500–after 432 B.C.)
Greek sculptor, anciently considered the greatest of all, though no work remains that can be certainly attributed to him. He was the general superintendent of work at the Parthenon for which he made a colossal statue of Athena.

Picasso, Pablo (b. 1881)
Sculpture has been part of the many-sided activities of the Spanish artist, in which he has been as various as in painting. His work includes modelled heads, constructions made out of rods of iron and the conversion of ready-made objects into other forms.

Pisano, Giovanni (*c* 1240–*c.* 1328) and **Niccolò** (1206–78)
Italian sculptors and architects who began the great development of sculpture in Italy, partly through a return to ancient standards. The father Niccolò was assisted by his son Giovanni. A principal work in which both had a part was the pulpit of the cathedral at Siena (Plate 34).

Pollaiuolo, Antonio (*c.* 1432–98)
Italian painter, sculptor and goldsmith. He assisted Ghiberti in modelling the gates of the Baptistery at Florence and produced original work in bronze. His "Hercules and Antaeus" (Bargello, Florence) shows, like his painting, his interest in anatomy and the rendering of muscular action.

Polykleitos of Argos (fifth century B.C.)
Greek sculptor, the contemporary of Phidias and Myron, especially admired for his statues of athletes. His work is known by copies, the "Doryphoros"—athlete with javelin—being a celebrated masterpiece. He also made a version of the "Wounded Amazon", a subject in which four Greek sculptors of the time competed.

Praxiteles (fourth century B.C.)
Greek sculptor noted for the grace and naturalness of pose in his statues. An original which has survived is the "Hermes" from Olympia, *c.* 350 B.C., and several works are related to it in style. Much admired in ancient times, his "Aphrodite of Cnidus" exists only in a copy.

Riccio (Andrea Briosco) (*c.* 1470–1532)
Italian sculptor of the Paduan School, a great master of the bronze statuette, treating classical subjects of myth with great imagination. A fine example is his "Pan" (Ashmolean Museum, Oxford) (Plate 49). He also applied sculptural ornament to objects of use, such as door-knockers, lamps, etc.

Richier, Germaine (1904–1959)
French woman sculptor, devoted to sinister and mysterious effect in figures often constructed of fine threads of plaster.

Riemenschneider, Tilman (1460–1531)
German sculptor, outstanding in works equally noted for their elaborate craftsmanship and feeling. Masterpieces are the carved altar group of the Assumption of the Creglingen altar (1501–5) and the

tomb of the Emperor Henry II and his wife in Bamberg Cathedral (1513).

Rodin, Auguste (1840–1917)
French sculptor, one of the greatest of any period in the rendering of movement and command of form. He studied under Barye and, after much routine labour in early years, reached success in Paris when about 37. A colossal output includes the monumental "Gate of Hell" inspired by Dante, the "Burghers of Calais", "The Thinker", "The Kiss", and (in his time sensational) the memorial to Victor Hugo, represented nude, and his Balzac in dressing-gown (Plate 55). Many small and splendid figure studies are also represented in the Rodin Museum, Paris.

Roubiliac, Louis François (1695–1762)
French-born sculptor, the pupil of Nicolas Coustou, who settled in England where he produced many outstanding portrait busts and monuments. His Nightingale Monument in Westminster Abbey (Plate 51) and the "Garrick" (Victoria and Albert Museum) are notable works.

Rude, François (1784–1855)
French sculptor, especially celebrated for the group in high-relief on the Arc de Triomphe, Paris, the "Departure of the Volunteers, 1792" a spirited translation of the Marseillaise into stone (1836).

Rysbrack, Michael (*c.* 1693–1770)
Flemish-born sculptor who settled in London where, like his contemporary Roubiliac, he produced many monuments and portrait busts. The monu-

ment to Sir Isaac Newton (1731) (Westminster Abbey), is one of his best. He excelled in portraiture.

Sansovino, Jacopo (1486–1570)
Italian sculptor and architect. A Florentine, the pupil of Andrea Sansovino (or Contucci), he went to Venice in 1527 where he was active and influential both in sculpture and architecture.

Scheemakers, Pieter (1691–1770)
Flemish-born sculptor who worked in London. Among a number of monuments is that of Shakespeare in Westminster Abbey.

Scopas (fourth century B.C.)
Greek sculptor, a contemporary of Praxiteles, author of reliefs at the Mausoleum of Halicarnassus.

Sluter, Claus (beginning of fifteenth century)
Flemish sculptor who worked at the court of the Dukes of Burgundy at Dijon and gave an individual force to late Gothic sculpture. Almost to be compared with Michelangelo's "Moses" is his "Moses" (1401–2) (Chartreuse de Champmol).

Stevens, Alfred (1818–75)
English sculptor, painter and designer, thoroughly, though belatedly, imbued with the Renaissance spirit. His masterpiece was the monument to Wellington in St Paul's Cathedral, unfinished at his death and completed in 1912.

Stone, Nicholas (1586–1647)
English sculptor and architect, noted for the Baroque porch of St Mary's, Oxford, and the gates of the Botanic Garden, Oxford. His monument to John Donne at Westminster showing him in a shroud

carried out Donne's instructions with macabre effect.

Stoss, Veit (early sixteenth century)
German sculptor in whose work the change from Gothic to Renaissance in South Germany can be seen. A masterpiece was his "Angelus" for S. Lorenz church, Nuremberg (1517–18), showing the Virgin and the Angel of the Annunciation within a wreath of roses, suspended from the ceiling.

Thorwaldsen, Bertel (1770–1844)
Danish sculptor who spent many years in Italy, where like Canova he was a principal figure in the Neo-Classic revival. His work was cold and imitative but in great demand throughout Europe in his own time. His statue of Lord Byron is at Trinity College, Cambridge.

Tinguely, Jean (b. 1925)
Swiss-born sculptor who has produced machine-driven "mobiles", in which "sculpture" so nearly imitates machinery as to be accompanied by appropriate noise.

Torel, William (late thirteenth century)
English sculptor and goldsmith. Works by him are the effigies of gilded bronze of Henry III and Eleanor of Castille (1291–2), in Westminster Abbey, both figures being nobly idealized (Plate 36).

Torrigiani, Pietro (1470–1522)
Italian sculptor who brought the Renaissance style to England in the tomb for Henry VII and his queen in Westminster Abbey (1519) (Plate 37). A

tomb for Henry VIII at Windsor was left unfinished and the bronze melted down during the Commonwealth period. Torrigiani left England for Spain where he died.

Vantongerloo, Georges (b. 1886)
Flemish-born sculptor and painter who joined the Dutch "De Stijl" movement in 1917 and applied its abstract geometrical principles to sculpture. In later works he excluded art movements from view and embarked on a quasi-scientific study of space and visual effect in works carried out in wire and polychrome plexiglass.

Verrocchio, (Andrea del Cione) (1435–88)
Italian sculptor, painter and goldsmith with a busy workshop in which his entirely personal output was small. His greatness as a sculptor, however, is affirmed by the equestrian statue of Bartolommeo Colleoni at Venice (Plate 39). "Jesus and the Doubting Thomas" is a fine work at S. Michele, Florence.

Vischer, Peter (1455–1529)
German sculptor who worked at Nuremberg. His masterpiece was the statue of "King Arthur" in armour at Innsbruck. His three sons were also sculptors.

Watts, George Frederick (1817–1904)
English painter and sculptor. His principal plastic work was the statue of "Physical Energy", (1904), set up in the Matoppo Hills in South Africa as a monument to Cecil Rhodes (replica in Kensington Gardens). Other works were his equestrian statue of

Hugh Lupus, his memorial to the Bishop of Lichfield, for Lichfield Cathedral, Lord Tennyson for Lincoln and his bust of Clytie.

Wotruba, Fritz (b. 1907)
Austrian sculptor whose figures carved in stone have a Cubist element of geometry. Both standing and reclining figures show a notable dignity.

Zadkine, Ossip (b. 1890)
Russian-born sculptor whose work shows a poetic imagination. Principal works include the "Homage to Rodin" (1945), the "Birth of Forms" (1947), "The Destroyed City" (Rotterdam memorial) (1951), the Van Gogh monument, Auvers (1956), and "Atomic Power" (1963).

INDEX

SCULPTURE — A.D. 50

DATE A.D.	ITALY & Eastern Europe	FRANCE	GERMA & Northern
500			
	Byzantine Sculpture		Norse & Ce Art
700		Early Christian	Sculpture
		CAROLINGIAN AGE	
900			
	Torcello		Hildeshei
	St. Mark's	ROMANESQUE AGE	
1100	Venice	France, Germany, Britai	
		GOTHIC AGE Sculpture of Chartres, Reims, W	
1300	THE RENAISSANCE The Pisani Orcagna della Quercia Ghiberti Donatello		Claus Slut Henry VII's Ch
1500	Verrocchio Michaelangelo	Goujon School of Fontainebleau	Veit Stoss Peter Visc Berruguet
	BAROQUE AGE Bernini	Coysevox	Baroque in
1700		Girardon	& S. Germ
	ROCOCO Classical Revival - Canova	Falconet Carpeaux Rodin	Flaxman Stevens
1900		Painter-Sculptors MODERN FORMS of Sculpture	Epstein Moore